Smart Structures

Smart structures and smart structural components have unusual abilities: they can sense a change in temperature, pressure, or strain; diagnose a problem; and initiate an appropriate action to preserve structural integrity while continuing to perform their intended functions. Smart structures can also store processes in memory and learn to repeat the actions taken. Among the varied applications of smart structures are aircraft sensors that warn of impending cracks and medical devices that monitor blood sugar and deliver insulin.

This text provides the basic information needed to analyze and design smart devices and structures. Among topics covered are piezoelectric crystals, shape memory alloys, electrorheological fluids, vibration absorbers, fiber optics, and mistuning. A final chapter offers an intriguing view of biomimetics and design strategies that can be incorporated at the microstructural level, deriving inspiration from biological structures.

The design of smart structures is on the cutting edge of engineering research and development, and there is a great need for an introductory book on the subject. This book will be welcomed by both students and practicing engineers.

A. V. Srinivasan is President of Strategic Technologies, Inc. and a Distinguished Visiting Professor at Worcester Polytechnic Institute.

D. Michael McFarland is the President of Caber Engineering in Glastonbury, Connecticut.

Smart Structures

Smart Structures

ANALYSIS AND DESIGN

A. V. SRINIVASAN
Strategic Technologies, Inc.

D. MICHAEL McFARLAND
Caber Engineering, Inc.

CAMBRIDGE UNIVERSITY PRESS
Cambridge, New York, Melbourne, Madrid, Cape Town, Singapore,
São Paulo, Delhi, Dubai, Tokyo, Mexico City

Cambridge University Press
32 Avenue of the Americas, New York NY 10013-2473, USA

www.cambridge.org
Information on this title: www.cambridge.org/9780521659772

First published 2001

A catalog record for this publication is available from the British Library

Library of Congress Cataloging in Publication data

Srinivasan, A. V

Smart structures : analysis and design / A. V. Srinivasan, D. Michael McFarland.

p. cm.

Includes bibliographical references and index.

ISBN 0-521-65026-7 – ISBN 0-521-65977-9 (pbk.)

1. Smart materials. 2. Adaptive control systems. 3. Piezoelectric devices.
4. Microactuators – Materials. 5. Shape memory alloys. 6. Biosensors. I. McFarland, D.
Michael, 1962– II. Title.

TA418.9.S62 S65 2000
620.1′1 – dc21 00-023038

ISBN 978-0-521-65026-7 Hardback
ISBN 978-0-521-65977-2 Paperback

Contents

Preface

Structural design in the traditional context once simply meant a selection of the dimensions of load-bearing components of a structure. Essentially two disciplines, mathematics and materials science, were integrated to obtain a structural design. Mathematics ensured equilibrium of the structure, and materials science provided material properties such an modulus of elasticity, yield stress and ultimate stress. A suitable material was chosen, and the design was considered complete upon establishing a factor of safety and a further check to ensure structural stability using a criterion such as Euler load to prevent buckling. Such an approach was considered adequate as long as the material selection was limited to wood and metals.

Advanced research in materials science resulted in man-made materials, such as plastics and composites. Selection of unusual shapes in the design of structural components and ideas of embedding sensors to monitor complex strain fields then took hold. Furthermore, materials with unusual properties were discovered: properties by which material behavior can be varied depending upon the phase of the material (e.g., shape memory alloys, such as NiTiNOL, whose phases change at critical temperatures), the poling direction (as in piezoelectric materials such as PZT), and the level of electric field (electrorheological fluids). These discoveries have opened up the design space to such an extent that possibilities of designing structures that can not only monitor themselves but also adapt to the environment are now contemplated by the research community.

This is the background that has ushered in an era of research efforts leading to "smartness" in structural design. As funds have become available to pursue research in this area, terminologies have been introduced to define the field of study. Not unexpectedly, a variety of names, such as smart materials, intelligent materials, and adaptive structures, have been proposed.

A certain amount of exaggeration is to be expected when a new field emerges, but to ascribe cognitive abilities to a structure would be, in our opinion, beyond the capabilities that can be integrated into a structural design. Similarly, we wish to emphasize that materials have properties that may be interesting and unusual, but smartness can be defined in the context of structures only when materials properties are exploited to serve a design function better than is possible through a conventional structural design.

It is with this conviction that we introduce the concept of a complete structure that can be the basis, in the new millennium, to design structural components. The viewpoint proposed in this book is that functions of sensing, diagnosing, and assessing the health of structures and actuating structures should all be integrated to form the complete structure, and this philosophy forms the basis for developing smart structures. It is important to note that smartness can be inherent in a structure by virtue of the material microstructure or the design itself. For example, a rope is a smart structure because of the hierarchical design in which individual wires are assembled to form a strand and several strands woven into a wide variety of rope arhitectures. The concept of hierarchy is incorporated into the design and provides the basis for smartness. Rope structures simulate a similar biological structure, tendon, whose architecture is highly hierarchical, with seven levels varying from atomic to microlevels. Thus, one is tempted to examine structural systems in nature that could provide broad guidelines for the structural engineering design of man-made components. Although it is futile to look for engineering analogs in the biological world, certain strategies developed by biological structures through evolution could be helpful in conceiving, designing, developing, manufacturing and testing smart structures.

This book is developed through courses taught first as a graduate seminar at the University of Connecticut in 1994 and 1995 and later as a regular course open to senior undergraduate and graduate students at Worcester Polytechnic Institute in 1996 and 1997. Also, a special week-long course was conducted in June of 1997 at the University of Kassel in Germany covering piezoelectrics, shape memory alloys, gyroscopic vibration absorbers, mistuning and biomimetics. In addition, the organization of the chapters draws upon the authors' research efforts in the areas of vibration absorption, multiplexing shape memory wires in composites to enhance frequency bandwidth of plate-like structures, fiber optics, mistuning and biomimetics. The example problems chosen are aimed at driving home the principles discussed in the text and provide an opportunity to the student to apply the same in typical engineering structural design situations.

We believe that the time is ripe for innovation by means of which smartness in structures can come about through development of new materials with required properties and new designs that allow integration of multiple functions of sensing, diagnosing, and actuating the structure, leading to vastly enhanced structural integrity. Innovations are therefore needed in a variety of engineering disciplines that cover the functions described above. Analysis procedures that permit modeling at the microstructural level are needed, as are instruments that sense and actuate at the level. Manufacturing processes that allow tailoring structural components from the micro to macro levels are needed. We hope that the topics presented in this text book serve to provide some inspiration to scientists and engineers in academia, government and industry.

The first author thanks (1) his wife Kamla for her continuing support, encouragement and patience during the preparation of this book, (2) his coauthor without whose keen interest and cooperation this effort would not have been possible and (3) Professor M. Noori for giving him the oppportunity to develop the course at WPI.

The second author wishes to thank the friends and family members whose sustained interest and encouragement were invaluable during the completion of this manuscript, and his coauthor for the example he has set through his scholarship and for the friendship that has made a sometimes arduous task much more rewarding than it would otherwise have been.

Many colleagues have helped in the course of preparing this textbook. Special thanks are due to Professor Dryver Huston of the University of Vermont for reviewing the chapter on fiber optics; to Dr. L. McD. Schetky and Dr. Ming Wu of Memry Corporation for help in obtaining permissions from Raychem Corporation to use several illustrations pertaining to the design of SMA actuators; to Professors B. F. Spencer and Erik A. Johnson of the University of Notre Dame for assistance with the chapter on electrorheological and magnetorheological fluids; to Professor Daniel J. Inman of the Virginia Polytechnic Institute, himself an authority in the field of smart structures, for his insights into the process of writing a textbook; and to Mr. Richard W. Monahan, whose patience and attention to detail have contributed in many ways to the quality of this book.

1

Introduction

A structure is an assembly that serves an engineering function. Examples of engineering structures can vary from a building or a bridge to a power plant or ship. Thus, helicopters, roads, and railways are structures or structural systems. The elements of these structures, such as wings of an airplane, blades of a helicopter, and blades in a jet engine, are also structures or structural components. Furthermore, in this context, microchips; semiconductors; measurement devices, such as accelerometers and strain gages; sports equipment, such as tennis rackets, and musical instruments; and cooking vessels may all be looked upon as structures or structural components serving engineering functions.

Structural design means selecting a profile, configuration, size, cross section, and material in order to meet the functions a structure has to perform (e.g., carry or transmit a load, perform work, or generate power).

Clearly, the dictionary definition of "smart" (brisk, spirited, mentally alert, bright, knowledgeable, shrewd, witty, clever, stylish, being a guided missile, operated by automation) is not quite adequate in this context. The engineering community has adapted the term *smart structures*, over nearly a decade now, and the words have come to mean a certain extraordinary ability of structures or structural components in performing their design function. Smartness, in this context, implies (a) the ability of structural members to sense, diagnose and actuate in order to perform their functions (closed-loop smartness) and/or (b) an unusual micro- or macro-structural design that enhances structural integrity (open-loop smartness). A closed-loop smart structure or component is one which has the ability to sense a variable such as temperature, pressure, strain, and so forth, to diagnose the nature and extent of any problem, to initiate an appropriate action to address the identified problem, and to store the processes in memory and "learn" to use the actions taken as a basis next time around. The attributes of smartness may thus include the abilities to self-diagnose, repair, recover, report, and learn. Examples of both types of smartness are shown in Figs. 1.1 through 1.5. One of the critical requirements in any structure is preservation and enhancement of its structural integrity, that is, its longevity or life through which it will continue to fulfill its design function. Any consideration of smartness in a structural design must have

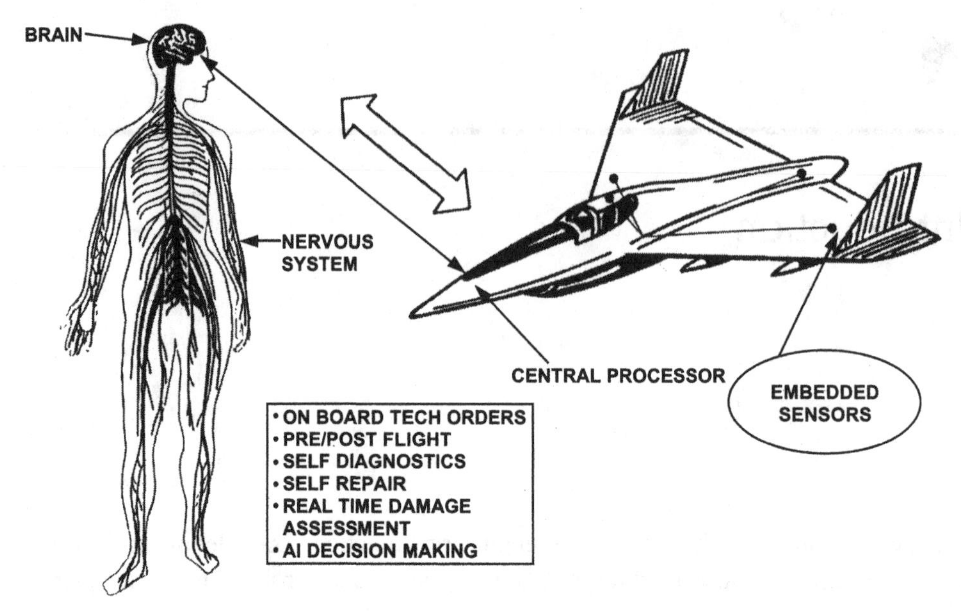

Figure 1.1. Schematic of smart aircraft response (closed-loop) (U.S. Air Force Office of Scientific Research).

this feature as a priority along with the necessary constraints of safety, affordability, reliability and cost-effectiveness. In the process of designing and building such a structure, it will be apparent that we need to embed sensors, processors and actuators to incorporate smartness to help preserve the integrity of the structure.

Materials are the elements, constituents, or substances of which something is composed or made. Examples include steel, wood, aluminum, concrete, composites, and plastics. Materials have properties; some of these may even be extraordinary properties. Materials with interesting, unusual, extraordinary and useful properties can be used to design and develop structures that can be called smart – but the material itself is not smart. Consider, for example, a material that may be stretched and bent at one temperature (like rubber) and retains that deformation upon removal of the loads (unlike rubber). That is its property. When the same material recovers its shape

Figure 1.2. The highly maneuverable dragon fly is a familiar example of a closed-loop smart structure.

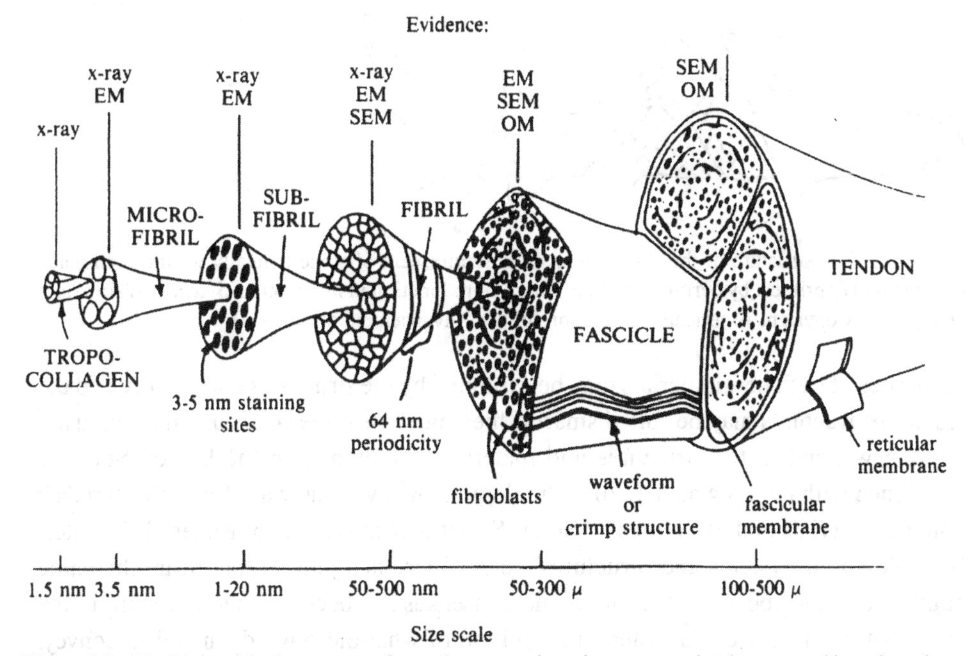

Figure 1.3. The hierarchical structure of a tendon comprises tissues organized on several length scales. (Courtesy of Eric Baer; Kastelic and Baer, 1980.)

completely at another temperature, then its property is interesting, even unusual. But the material is not smart, although we can use it as a structural component (a fiber or a spring) to accomplish a desired design function. For example, if we were to use this interesting material property to develop a spring supporting a heavy load and to lift the load periodically by varying the temperature of the spring, then we may call the resulting structure smart if, and only if, the resulting system is better than any other conventional method to lift loads. Thus, unusual, interesting, and even amazing

Figure 1.4. The shell of a mollusk exhibits high fracture toughness because cracks cannot propagate through it in a straight path (Pacific Northwest Laboratory, U.S. Department of Energy, 1988).

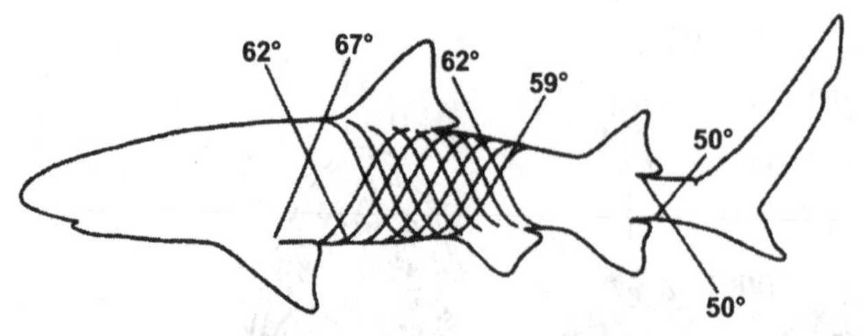

Figure 1.5. The structure of the skin of a shark is reminiscent of some helically wound man-made composites (Reprinted, by permission, from Shark skin: function in locomotion, *Science*, Wainwright et al., 1978. Copyright American Association for the Advancement of Science).

properties of such new materials may be exploited by the structures community and the resulting structures may be called smart if they meet the criteria discussed previously.

A new trend in the structures and materials community in the United States is concerned with defining an area of technology known variously as "Smart Materials"; "Smart Materials and Structures"; "Smart Structures"; "Intelligent Materials"; "Intelligent Material Systems and Structures"; etc. (There is a journal that bears the name at the end of the above list.) Although no consensus has been reached in regard to the choice of words, there is a general recognition of what these words intend to convey. Consider, for example, characterizations such as the following: performing sensing and actuation functions, recognizing change and initiating an appropriate response, having the ability to learn, and being capable of adapting to external stimuli.

A commonly accepted hierarchy of abilities in humans reaches the highest attribute of wise, followed by intelligent, smart, sensible, trivial, foolish, and dumb, and ends with stupid. Thus, it should be clear that it is inappropriate to use the word "smart" to define materials. It is even worse to use the word intelligent in order to describe materials, because the word intelligent connotes cognitive capabilities and, based on the above discussion, it is imprecise to use that word to define even structures, let alone materials. Intelligence implies capabilities far superior to what may be possible even in a smart structure. Take for example the action of slewing a human arm. When you move your arm from left to right quickly, you can stop when you want to and you do not notice any vibration of your arm. Think of simulating this action by using a beam. It requires many sensors and actuators to accomplish the simple act of slewing without causing vibration. Slewing your arm in a natural act is controlled so very precisely, with an enormous number of precisely coordinated muscular actions, that duplication of that motion in a structural equivalent can be accomplished only in a limited way. The structural arm may be smart but it cannot be intelligent, simply because it is impossible to simulate the billions of micro-actions taking place in the human arm controlled by the yet-to-be-understood controller par excellence called the brain.

Smart structural design means a consideration of functions a structure needs to perform, under the influence of an environment, in order to select and balance a material organization with the selection of the cross section, profile, or size of a structure or structural component into which are incorporated instruments that have the ability to diagnose, assess, and initiate an action appropriate to meet a design intent. This defines what we call a *complete structure* whose design foresees the final product in all its detail and avoids sequential development. Examples of the opposite type are abundant. Familiar scenes in modern cities are roadways that get dug up *after finishing* in order to accommodate or relocate water pipes, gas pipes, and telephone lines. Another example is composite panels that are drilled to make holes *after* they are cured.

We may conclude by attempting a definition of smart structures as follows: Smart structures have the capability to sense, measure, process, and diagnose at critical locations any change in selected variables, and to command appropriate action to preserve structural integrity and continue to perform the intended functions. The variables may include deformation, temperature, pressure, and changes in state and phase, and may be optical, electrical, magnetic, chemical, or biological. The question of structural integrity arises when defects develop, cracks form and propagate, or vibration occurs at resonance or flutter. Some examples are earthquake response of buildings, cutting tool chatter, rotor critical speeds, and turbine engine blade flutter.

The subject of smart structures is interdisciplinary, encompassing a variety of subjects including materials science (metallurgy, composites), applied mechanics (vibrations, fracture mechanics, elasticity, aerodynamics), electronics (sensors, actuators, controls), photonics (fiber optics), manufacturing (processing, microstructure), and biomimetics (strategies adopted by natural structures).

Some examples of potential smart structural systems and some mechanisms that are candidates for smart structure application are listed below:

Aircraft: Monitoring at key locations on the aircraft the state of strains to warn the pilot of any impending development and propagation of cracks; wings that alter shape with respect to air speed and pressure to increase performance and fuel efficiency (Fig. 1.1).

Spacecraft: Pointing accuracy of large antennas maintained through an elaborate network of sensors and actuators.

Buildings: Buildings designed to resist earthquake damage; smart windows: electrochromic windows that sense weather changes and human activity and automatically adjust light and heat.

Bridges: Remote monitoring of strains, deflections, and vibration characteristics in order to warn of impending failures.

Ships: Hulls and propulsion systems that detect noise, remove turbulence and prevent detection.

Machinery: Tool chatter suppression; rotor critical speed control.

Jet engines: Fan, compressor, and turbine blades that exploit asymmetry arising out of nonuniformities in structural and/or aerodynamic properties.

Pipelines: Monitoring of leakage and damage to underground pipes for water, oil, and gas.

Medical devices: Blood sugar sensors, insulin delivery pumps, micromotor capsules that unclog arteries. Filters that expand after insertion into a vessel to trap blood clots.

Smartness in structures is evident when toughness is enhanced through hierarchical design, such as tendon or rope-like structures; when cracks are arrested through mechanisms commonly found in sea shells; when natural phenomena, such as friction or inevitable nonuniformities in nominally uniform structures, are exploited; when shape changes are introduced in components by using phase change properties of materials; when optic fibers are used not only to sense a variable but also to carry loads; and when components are integrated to develop a monolithic part, such as a blisk.

In summary, we may state that structures are engineered to perform functions of carrying loads, temperature, and pressure, and conveying electrical, magnetic, and optical signals without experiencing irreparable damage during the service life for which they were designed. An engineered smart structure must meet the following four criteria: functionality, durability, affordability, and safety. More often than not, functionality is multiple, as in the case of aircraft wings carrying fuel in addition to developing lift forces, or tree roots carrying nutrients in addition to anchoring the tree to the ground.

Some materials of interest in developing smart structures are piezoelectric crystals, electrostrictive and magnetostrictive materials, shape memory alloys, electrorheological fluids, and fiber optics. In the following chapters, basic characteristics of some of these materials are discussed in the context of their potential to develop smart structures.

BIBLIOGRAPHY

Amato, I. 1990. Smart as a brick. *Science News* 137(10):152–153, 157.

Banks, H. T., R. C. Smith, and Y. Wang. 1996. *Smart Material Structures: Modeling, Estimation and Control.* New York: John Wiley & Sons.

Culshaw B. 1996. *Smart Structures and Materials.* Boston: Artech House.

Editorial. 1993. Intelligent material systems—The dawn of a new materials age. *Journal of Intelligent Material Systems and Structures* 4(1):4–12.

Gordon, J. E. 1976. *The New Science of Strong Materials: Or Why You Don't Fall through the Floor.* 2d ed. Princeton, New Jersey: Princeton University Press.

Srinivasan, A. V. 1996. Smart biological systems as models for engineered structures. *Materials Science & Engineering* C(4):19–26.

2

Piezoelectric Materials and Induced-Strain Actuation

2.1 Introduction

Certain materials possess a property by which they experience a dimensional change when an electrical voltage is applied to them. Such materials are known as *piezoelectric* because of the converse effect; that is, they generate electricity when pressure is applied. Perhaps the best-known such material is lead-zirconate-titanate (PZT); in fact, "PZT" is commonly used to refer to piezoelectric materials in general, including those of other compositions.

2.2 Piezoelectric Properties

When manufactured, a piezoelectric material has electric dipoles arranged in random directions. The responses of these dipoles to an externally applied electric field would tend to cancel one another, producing no gross change in dimensions of the PZT specimen. In order to obtain a useful macroscopic response, the dipoles are permanently aligned with one another through a process called *poling*.

A piezoelectric material has a characteristic *Curie temperature*. When it is heated above this temperature, the dipoles can change their orientation in the solid phase material. In poling, the material is heated above its Curie temperature and a strong electric field is applied. The direction of this field is the polarization direction, and the dipoles shift into alignment with it. The material is then cooled below its Curie temperature while the poling field is maintained, with the result that the alignment of the dipoles is permanently fixed. The material is then said to be poled.

When the poled ceramic is maintained below its Curie temperature and is subjected to a small electric field (compared to that used in poling), the dipoles respond collectively to produce a macroscopic expansion along the poling axis and contraction perpendicular to it (or vice versa, depending on the sign of the applied field). The geometry and deformation of a simple cube of PZT, which has been poled in the

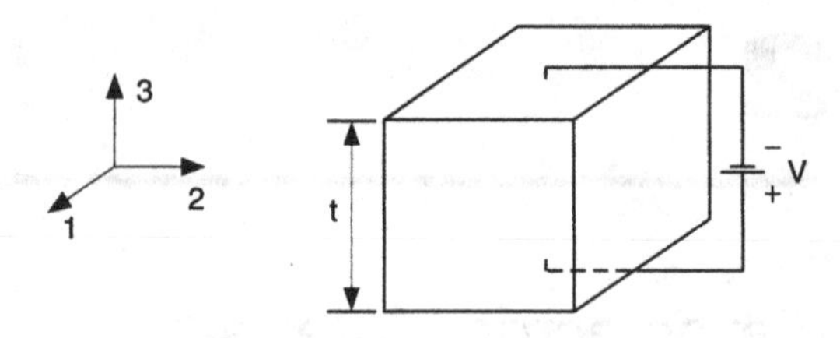

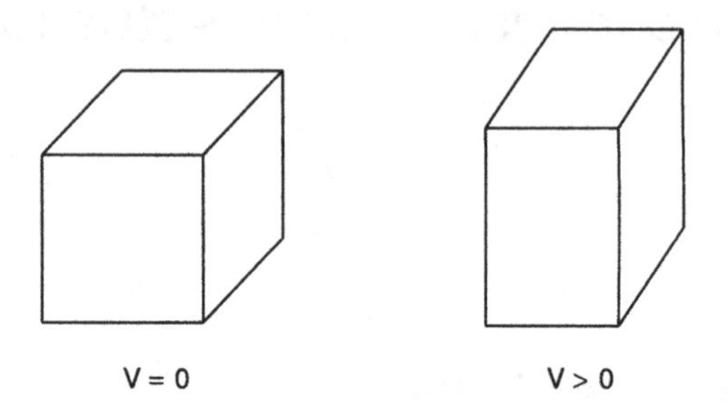

Figure 2.1. Deformation of a cube of PZT subjected to a uniform electric field.

3-direction and is then subjected to an electric field in this direction, is shown in Fig. 2.1.

The relationship between applied field strength and resulting strain is quantified by the *piezoelectric moduli* d_{ij}, where i is the direction of the electric field and j the direction of the resulting normal strain. Thus, for example,

$$\varepsilon_{33} = d_{33}\frac{V}{t} \tag{2.1}$$

and

$$\varepsilon_{11} = d_{31}\frac{V}{t} \tag{2.2}$$

where V is the voltage applied in the 3-direction and t the thickness of the specimen, as shown in the figure. Typical values of the piezoelectric moduli are given in Table 2.1. Note that for the same applied voltage (field strength), the soft PZT will experience a greater deformation.

The working temperature of the PZT is usually well below its Curie temperature. If the material is heated above its Curie temperature when no electric field is applied, the dipoles will revert to random orientations. Even at lower temperatures, the application of too strong a field can cause the dipoles to shift out of the preferred alignment established during poling. Once depoled, the piezoelectric material loses the property of dimensional response to an electric field.

Table 2.1. Typical Piezoelectric Moduli of PZT Materials (m/V)

	d_{33}	d_{31}
Hard PZT	225×10^{-12}	-100×10^{-12}
Soft PZT	600×10^{-12}	-275×10^{-12}

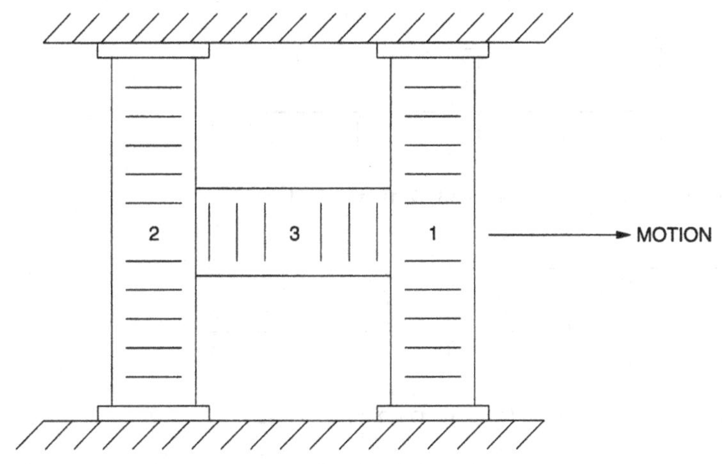

Figure 2.2. Inchworm motor made of 3 PZT stacks.

Hard PZT materials have Curie temperatures above 300 °C and are not easily poled except at elevated temperatures. Soft PZT materials have Curie temperatures below 200 °C and are readily poled or depoled at room temperature with strong electric fields.

2.2.1 Example: Inchworm Linear Motor

Using only the properties discussed so far, it is possible to describe the operation of a realistic piezoelectric mechanism. Shown schematically in Fig. 2.2 is an "inchworm" type linear motor consisting of an H made of three piezoelectric stacks. Each stack is made up of several PZT elements that expand or contract in response to an applied voltage to produce axial motion of the ends of the stacks. In this configuration, stacks 1 and 2 can be expanded to clamp against the walls of the channel, while stack 3 is alternately expanded and contracted.

By properly synchronizing the voltages applied to the PZT stacks, the H may be made to move in either direction along the channel. To produce motion in the direction indicated, signals as shown in Fig. 2.3 may be used, causing the armature of the motor to advance in "steps" as shown in Fig. 2.4. The velocity that can be achieved depends on the rate at which the control electronics can generate the needed signals and on the axial displacement of the motor achieved in each step. For example, suppose each PZT stack is 1 cm in length and made of a material with piezoelectric modulus 600×10^{-12} m/V, and that the controller can generate 2 000 output changes per second at a voltage sufficient to produce a field of ± 500 kV/m in the actuators. From the timing diagram it will be seen that each step through the full range of stack 3 will require six controller actions, so the motor armature can advance at the rate of

$$\frac{2\,000}{6} = 333.3\,\frac{\text{steps}}{\text{s}} \tag{2.3}$$

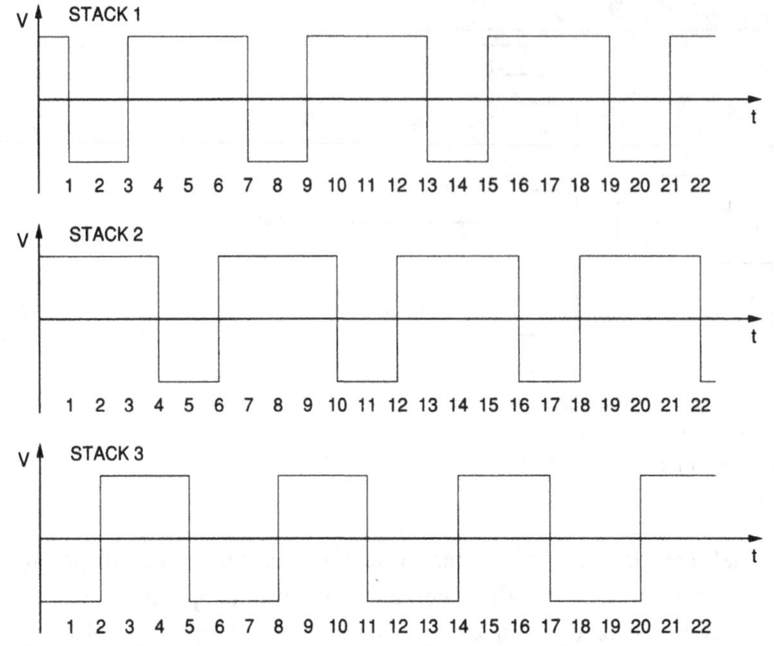

Figure 2.3. Timing of signals driving the inchworm motor PZT stacks.

The maximum strain developed in actuator 3 will be

$$\varepsilon = \left(500 \times 10^3 \,\frac{\text{V}}{\text{m}}\right)\left(600 \times 10^{-12}\,\frac{\text{m}}{\text{V}}\right) = 300 \times 10^{-6} \tag{2.4}$$

and the corresponding displacement over the 1 cm length of the PZT stack will be

$$\Delta = (300 \times 10^{-6})(0.01\,\text{m}) = 3\,\mu\text{m}. \tag{2.5}$$

Because the applied voltage will vary from positive to negative, each step will advance the motor:

$$1\ \text{step} = 2\Delta = 6\,\mu\text{m}. \tag{2.6}$$

Therefore,

$$\text{velocity} = \left(\frac{6\,\mu\text{m}}{\text{step}}\right)\left(\frac{2\,000}{6}\,\frac{\text{step}}{\text{s}}\right) = 2\,000\,\frac{\mu\text{m}}{\text{s}} = 2\,\frac{\text{mm}}{\text{s}}. \tag{2.7}$$

2.3 Actuation of Structural Components by Piezoelectric Crystals

In most applications where it is desired to drive a structure by using one or more piezoelectric actuators, the PZT elements are bonded to or embedded in a passive base structure. Here we will consider one-dimensional structures (rods and beams) to which are attached PZT actuators. In each case, we will assume a perfect bond of the actuator to the structure, so that displacement is continuous at their interface, and we

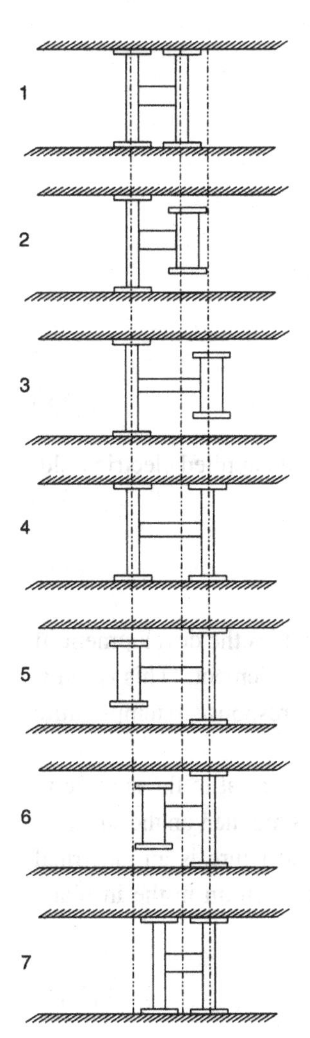

Figure 2.4. Sequence of PZT stack actions to advance the inchworm motor armature by one step.

will make the conventional assumptions about stress and strain distributions with such modifications as are necessary to take account of the electrically induced strain in the PZT. The resulting external load term in the structures' equations of motion will be seen to follow naturally from this analysis.

2.3.1 Actuator-Structure Interaction

It is important to realize that when a piezoelectric element is used in this way, the strain ε_a in the PZT actuator almost always is the result of the superposition of two components: the "free strain" ("piezoelectric strain") ε_p, which would result were the same voltage applied to the PZT element alone, and a mechanical strain arising from load produced on the PZT because of deformation of the base structure to which it was attached. These strain components will necessarily be of opposite signs if the actuator

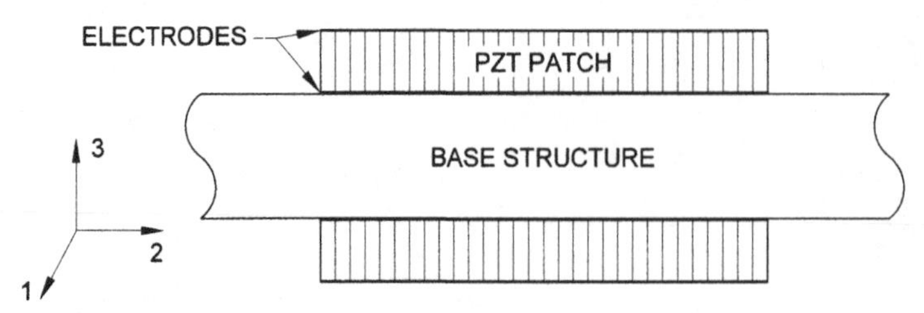

Figure 2.5. Typical geometry of PZT patches bonded to a base structure.

is to develop a force on the structure, implying that

$$|\varepsilon_p| > |\varepsilon_a|. \tag{2.8}$$

The free strain depends only upon the piezoelectric modulus and applied electric field, and is always given by

$$\varepsilon_{\bar{j}\bar{j}} = d_{ij}\frac{V}{t}. \tag{2.9}$$

The name "free strain" reflects the fact that no stress accompanies the development of piezoelectrically induced strain in an unconstrained (free) PZT element. (This is reminiscent of the stress-free thermal strains occurring in metals in response to temperature changes.)

In many applications the poling axis of the PZT patch is normal to the surface to which the patch is bonded, as shown in Fig. 2.5. The electrodes are then on the surfaces of the PZT parallel to the surface of the structure. (If the structure is an electrical conductor, it may be used as one side of the circuit.) The free strain in the in-plane directions is then

$$\varepsilon_{11} = \varepsilon_{22} = d_{31}\frac{V}{t}. \tag{2.10}$$

In the discussion to follow, we shall assume that the 1-axis of the piezoelectric actuator is aligned with the 1- or x-axis of the base structure, so that the free strain of the PZT is given by eq. (2.10). The strain in the base structure, which has only a mechanical component, will be denoted by ε_s. Additional subscripts will be used as necessary to indicate particular components of strain or strain at the interface of the structure and the actuator. The Young's moduli of the base structure and of the piezoelectric actuator will be denoted by E_s and E_a, respectively.

The assumption of a perfect bond implies continuity of displacement at the interface of the structure and the actuator. Ordinarily, this would in turn imply continuity of strain, and indeed if the structure is deformed solely by external mechanical loading, the resulting strains will be continuous. However, when an electric field is applied to the PZT it will develop an additional strain that will be superposed on any mechanically induced strain, resulting in a discontinuity in strain at the actuator-structure interface. The magnitude of this discontinuity will be exactly equal to that of the free strain ε_p, and is independent of any mechanical loading or deformation.

The signs of the mechanical and piezoelectric components of strain may or may not be the same. In the most common case, a voltage is applied to a PZT actuator with the intention of transferring load to the base structure. Suppose the structure is otherwise unloaded and the applied voltage would produce a tensile free strain were the actuator unconstrained. Because it is attached to the structure, the PZT cannot expand freely, but rather is constrained to undergo the same displacement as the structure. The result is that the structure is subjected to tractions in the direction of expansion of the patch, while the PZT, because is constrained, experiences a *compressive* stress. Therefore, the net strains in the structure and actuator are of opposite signs, and are limited by the piezoelectric strain ε_p.

We will now examine these considerations in more detail for the cases of axial and bending deformation.

2.3.2 Axial Motion of Rods

We will first consider the piezoelectrically induced axial displacement of a rod. As shown in Fig. 2.6, the system consists of a flat bar to which are attached two PZT patches. These are attached to the controller circuitry so that they expand or contract together, and hence the induced strains and stresses can be expected to be symmetric about the rod's midplane.

The rod may be loaded by an external force F, by the PZT actuators, or by both. Below we calculate the strain and stress distributions through the rod and PZT for each of these loading cases.

1. *Mechanical Loading Only ($V = 0$, $F > 0$)*
   Strain is uniform through the section, as shown in Fig. 2.7. The stresses differ in the materials because their elastic moduli are unequal. Assume $E_a > E_s$; the stress distribution will then be as shown in Fig. 2.8, with

$$\sigma_s = E_s \varepsilon_s \tag{2.11}$$

and

$$\sigma_a = E_a \varepsilon_s, \tag{2.12}$$

where ε_s is the strain in the base structure (rod) and, in this case, in the PZT as well.

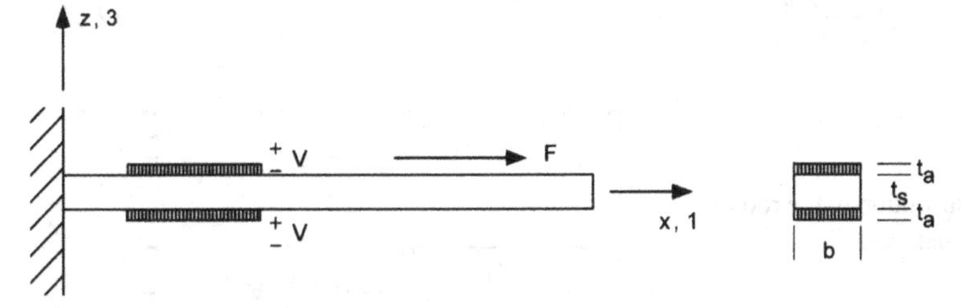

Figure 2.6. Rod with symmetric PZT actuator patches.

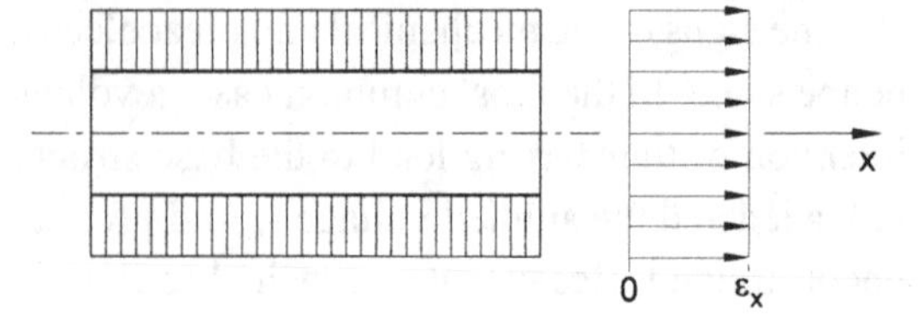

Figure 2.7. Strain distribution in the rod and PZT due to the external load F.

Force equilibrium at this section requires

$$2\sigma_a t_a b + \sigma_s t_s b = F. \tag{2.13}$$

Substituting eqs. (2.11) and (2.12) for the stresses leads to

$$\varepsilon_x = \frac{F/b}{2E_a t_a + E_s t_s}. \tag{2.14}$$

2. *Actuator Loading Only ($V > 0$, $F = 0$)*
As before, displacement is uniform through the section (this follows from the usual assumptions made in modeling a slender rod), but now the strains differ in the two materials. This difference is the electrically induced free strain ε_p. Let the polarity of the applied voltage be such as to cause extension of the actuators in the x-direction. This will stretch that portion of the rod that lies between the PZT patches, and hence $\varepsilon_s > 0$ in this region. (Of course, $\varepsilon_s = 0$ elsewhere in the rod.)

The actuators experience a strain equal to that in the rod plus the free strain ε_p. The superposition of these two components is shown in Fig. 2.9. If the actuators were completely constrained they would undergo a mechanical *compressive* strain of $-\varepsilon_p$. Here, because the rod does deform, they experience a net strain of $\varepsilon_s - \varepsilon_p$. The resulting stresses are

$$\sigma_s = E_s \varepsilon_s, \tag{2.15}$$
$$\sigma_a = -E_a(\varepsilon_s - \varepsilon_p). \tag{2.16}$$

Because $F = 0$, force balance may be stated:

$$2\sigma_a t_a b + \sigma_s t_s b = 0. \tag{2.17}$$

Consequently,

$$\varepsilon_s = \frac{2E_a \varepsilon_p t_a}{2E_a t_a + E_s t_s} \tag{2.18}$$

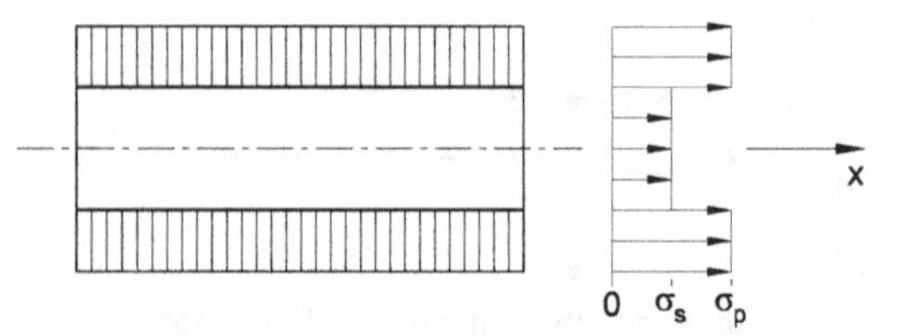

Figure 2.8. Stress distribution in the rod resulting from the external load F.

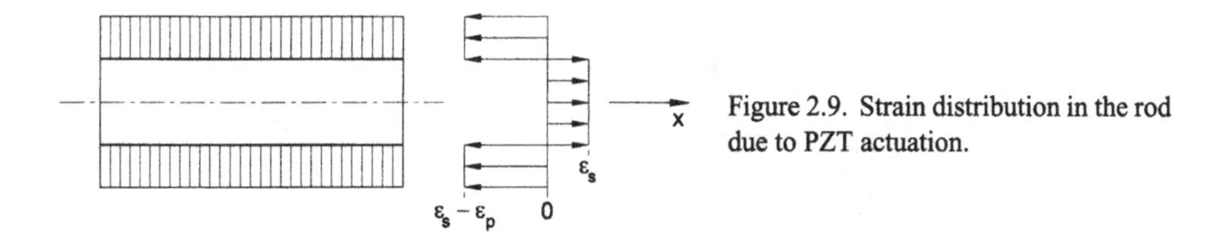

Figure 2.9. Strain distribution in the rod due to PZT actuation.

where

$$\varepsilon_p = d_{31}\frac{V}{t_a}. \tag{2.19}$$

3. *Simultaneous Mechanical and Piezoelectric Loading ($V > 0$ and $F > 0$)*
Linear behavior has been assumed in all of the above calculations, and so the response in this case can be found by superposition. Within the region where the actuators are attached, the axial strain in the rod is

$$\varepsilon_s = \frac{(F/b) + 2E_a\varepsilon_p t_a}{2E_a t_a + E_s t_s}. \tag{2.20}$$

2.3.3 Bending of Beams

The system to be considered here is very similar to the rod previously discussed, with two important differences: the PZT actuators bonded to the top and bottom surfaces of the structure are driven by voltages of opposite polarity, so that when one is expanded the other is contracted, and the resulting deformation of the base structure is modeled as bending rather than extension. The beam and piezoelectric actuators are shown in Fig. 2.10. A Euler-Bernoulli beam model is adopted, with the usual assumption that transverse plane sections remain plane during deformation and therefore stress and strain vary linearly through the thickness of the beam, except for discontinuities that may arise due to piezoelectric free strain.

If the beam and actuators were bent by an external (mechanical) load into an upward curvature, the portion of the beam above the neutral axis and the top PZT element would be placed in compression, and the bottom half of the structure in tension. The corresponding strain and stress distributions are shown in Figs. 2.11 and 2.12. Strain is continuous at the beam-actuator interface but stress is discontinuous because of the difference in the Young's moduli E_s and E_a of the beam and actuator materials.

In the following analysis, it will frequently be necessary to identify a quantity or value as occurring at the interface. Such terms will be given an additional subscript, h, in keeping with the notation of Fig. 2.10, where h is the distance from the neutral axis

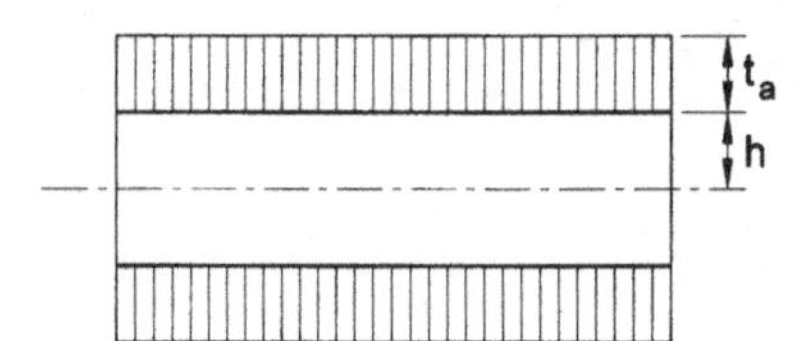

Figure 2.10. Geometry of beam and PZT actuators.

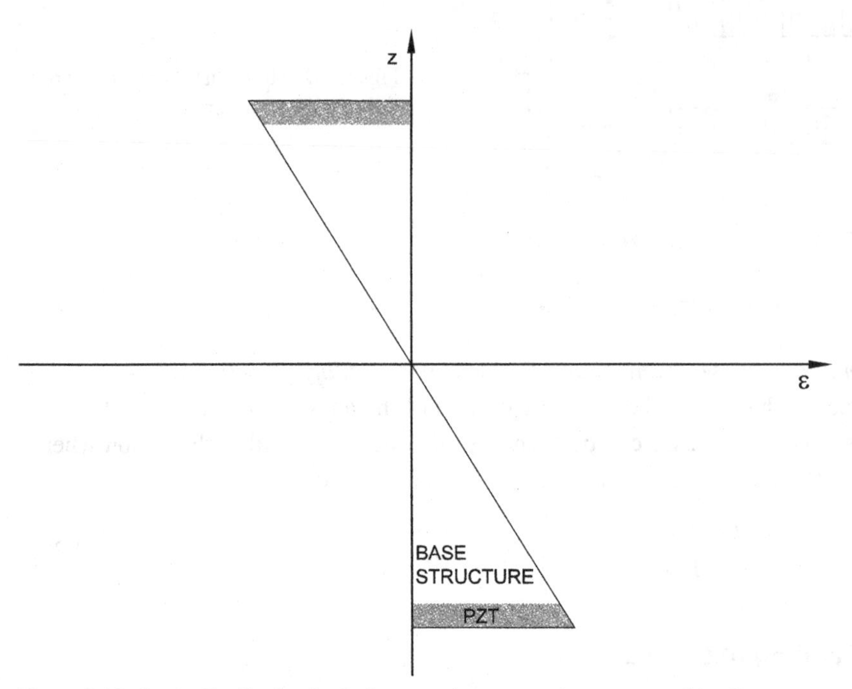

Figure 2.11. Strain distribution in the beam and actuators due to external loading.

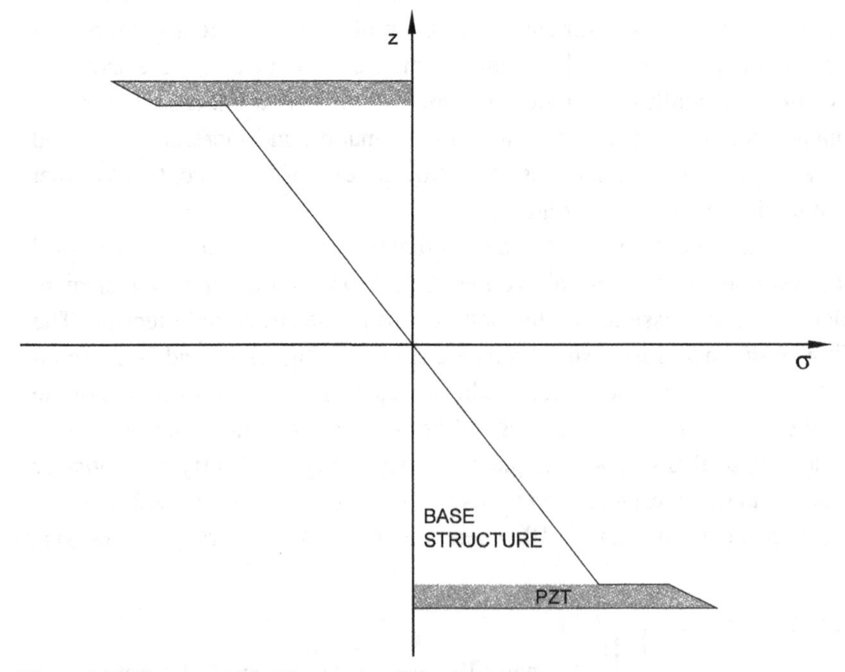

Figure 2.12. Stress resulting from external loading.

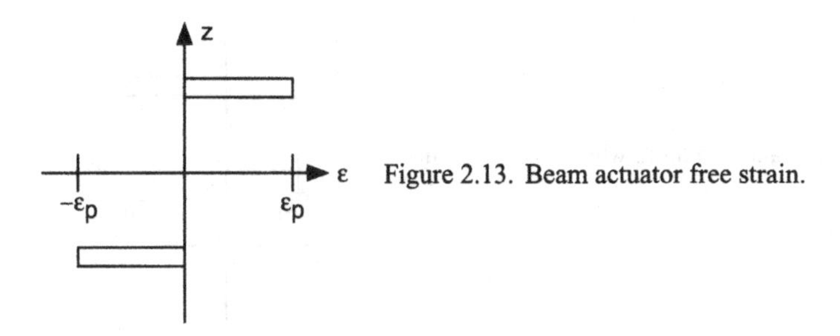

Figure 2.13. Beam actuator free strain.

to the top surface of the beam. Because the structure is symmetric about its midplane, $h = t_s/2$. In this notation, the mechanical load results in

$$\varepsilon_{a_h} = \varepsilon_{s_h}, \tag{2.21}$$
$$\sigma_{a_h} > \sigma_{s_h}, \tag{2.22}$$

where we've again assumed $E_a > E_s$.

Because the free strain in the PZT is independent of bending, it is constant through the actuator thickness, as shown in Fig. 2.13. Returning to the usual situation where bending of the beam is to be driven by the PZT actuators, the strain that produces stress in the PZT is the superposition of the free strain and the bending strain. This is shown in Fig. 2.14, where

$$\varepsilon_{a_h} = \varepsilon_{s_h} - \varepsilon_p \tag{2.23}$$

and ε_p is negative in the top actuator. The corresponding stress distribution is then obtained by multiplying the strain in each material by the appropriate elastic modulus, with the result shown in Fig. 2.15. In this way the PZT crystals exert a distributed moment on the beam. Our next goal is to find a relationship between this moment per unit length and the actuator free strain.

The linear variation of stress and strain within the beam may be expressed as

$$\varepsilon_s = \varepsilon_{s_h}\frac{z}{h}, \tag{2.24}$$

$$\sigma_s = E_s\varepsilon_s = E_s\varepsilon_{s_h}\frac{z}{h} = \sigma_{s_h}\frac{z}{h}. \tag{2.25}$$

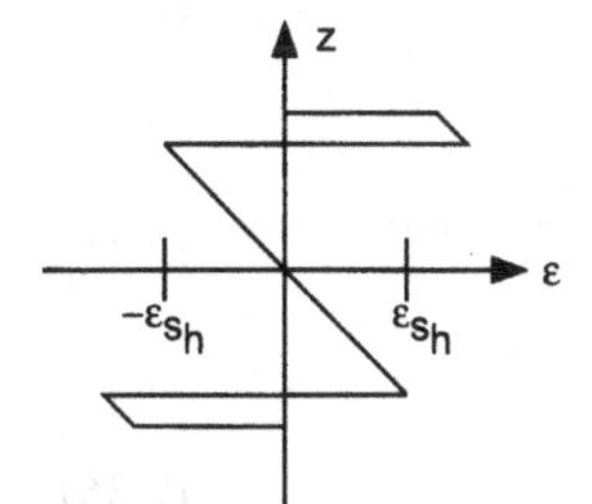

Figure 2.14. Distribution of stress-producing strain in the beam and actuators.

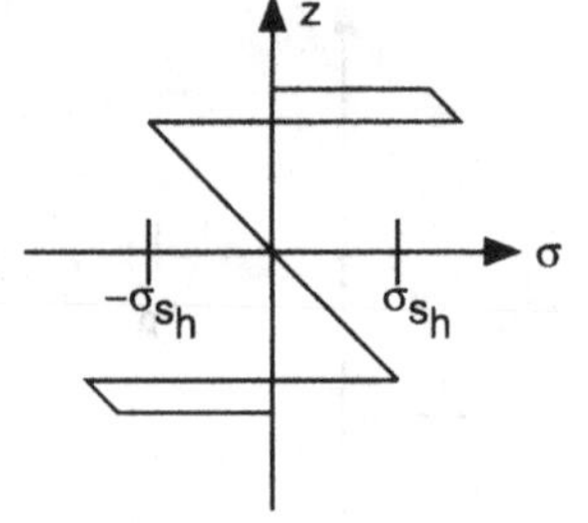

Figure 2.15. Stress in the beam and actuators when bending is induced by the PZT elements.

Likewise, in the crystal,

$$\varepsilon_a = \varepsilon_{s_h}\frac{z}{h} - \varepsilon_p, \tag{2.26}$$

$$\sigma_a = E_a\varepsilon_a = E_a\left(\varepsilon_{s_h}\frac{z}{h} - \varepsilon_p\right). \tag{2.27}$$

Setting $z = h$ yields

$$E_a\varepsilon_p = E_a\varepsilon_{s_h} - \sigma_{a_h}. \tag{2.28}$$

Then

$$\sigma_a = E_a\varepsilon_{s_h}\frac{z}{h} - E_a\varepsilon_{s_h} + \sigma_{a_h} \tag{2.29}$$

$$= \sigma_{a_h} - E_a\varepsilon_{s_h}\left(1 - \frac{z}{h}\right). \tag{2.30}$$

The above equation represents a relationship between stress within the PZT, stress in the beam at the interface, and piezoelectric strain.

The moments about the neutral axis due to the stress resultants must balance; thus

$$\int_0^h \sigma_s bz\,dz + \int_h^{h+t_a} \sigma_a bz\,dz = 0. \tag{2.31}$$

Substituting the expression developed above and integrating leads to the relationship between the stresses in the beam and the PZT at the interface. This may be written as

$$\sigma_{s_h} = K\sigma_{a_h} \tag{2.32}$$

where

$$K = \frac{-3ht_a(t_a + 2h)}{2\left(h^3 + E_R t_a^3\right) + 3E_R h t_a^2} \tag{2.33}$$

is a dimensionless number and

$$E_R = \frac{E_a}{E_s}. \tag{2.34}$$

Applying Hooke's law at the interface gives

$$\sigma_{s_h} = E_s\varepsilon_{s_h}, \tag{2.35}$$

$$\sigma_{a_h} = E_a\left(\varepsilon_{s_h} - \varepsilon_p\right), \tag{2.36}$$

and substituting these equations into the relationship just derived leads to an expression relating the strain at the interface to the actuator free strain:

$$\varepsilon_{s_h} = \frac{-P}{1-P}\varepsilon_p \tag{2.37}$$

where

$$P = KE_R. \tag{2.38}$$

Thus, the strain in the beam at the interface is a function of the free strain ε_p of the actuator, the ratio of the beam and actuator elastic moduli, and geometric parameters.

We may now consider bending caused by the actuator-induced stress in the beam. Integrating the triangular stress distribution in the beam gives immediately

$$m = \frac{2}{3}bh^2\sigma_{s_h} \tag{2.39}$$

and substituting for the stress produces

$$m = \frac{2}{3}bh^2 E_s \varepsilon_{s_h} \tag{2.40}$$

$$= -E_s \frac{P}{1-P}\frac{2}{3}bh^2\varepsilon_p. \tag{2.41}$$

This is the relationship between moment intensity and piezoelectric strain that was sought.

Having related the moment per unit length with the free strain in the PZT and the material and geometric properties, we will now compute the response of the beam to this moment loading. Central to this is the observation that the uniform distributed moment acting over the region $x_1 < x < x_2$ can be replaced in the beam's differential equation by two point moments, at x_1 and x_2.

The equation governing a static beam may be written

$$\frac{d^2}{dx^2}[M(x) - m(x)] = 0 \tag{2.42}$$

where $M(x)$ is the beam internal bending moment, and $m(x)$ is the externally applied moment intensity – in this case, the moment created by the PZT actuators. From the foregoing analysis, this external moment is known to be

$$m(x) = \begin{cases} C_0\varepsilon_p & \text{if } x_1 < x < x_2, \\ 0 & \text{otherwise,} \end{cases} \tag{2.43}$$

where

$$C_0 = -E_s \frac{P}{1-P}\frac{2}{3}bh^2. \tag{2.44}$$

This may also be written in terms of the Heaviside step function

$$H(x) = \begin{cases} 0 & \text{if } x < 0, \\ 1 & \text{if } x > 0 \end{cases} \tag{2.45}$$

as

$$m(x) = C_0\varepsilon_p[\cdot H(x - x_1) - H(x - x_2)]. \tag{2.46}$$

The first and second derivatives of the step function are $\delta(x)$ and $\delta'(x)$, which in this context may be interpreted as representing a point load and point moment, respectively. Thus,

$$\frac{d^2}{dx^2}m(x) = C_0\varepsilon_p\frac{d^2}{dx^2}[H(x - x_1) - H(x - x_2)] \tag{2.47}$$

$$= C_0\varepsilon_p\frac{d}{dx}[\delta(x - x_1) - \delta(x - x_2)] \tag{2.48}$$

$$= C_0\varepsilon_p[\delta'(x - x_1) - \delta'(x - x_2)]. \tag{2.49}$$

This represents two point moments of opposite sense, one acting at $x = x_1$, and the other at $x = x_2$. In other words, the moment distribution produced by the PZT actuators is equivalent to these point moments in respect to the response it produces in the beam.

Substituting $m(x)$ from above and

$$M(x) = EI\frac{d^2}{dx^2}w(x), \tag{2.50}$$

where $w(x)$ is the transverse displacement of the beam, into eq. (2.42) produces the fourth-order differential equation:

$$\frac{d^4}{dx^4}w(x) = C_0\varepsilon_p[\delta'(x - x_1) - \delta'(x - x_2)]. \tag{2.51}$$

This is perhaps the most familiar form of the beam's governing equation, and makes explicit the effective transverse load produced by the piezoelectric actuator patches.

2.3.3.1 Example: Harmonic Excitation

If the voltage applied to the actuators is time varying, so will be the moment and the equivalent force acting on the beam. The equation of motion should then include a dynamic term, and becomes

$$\frac{\partial^2}{\partial x^2}[M(x) - m(x)] + \rho A\frac{\partial^2}{\partial t^2}w(x,t) = 0. \tag{2.52}$$

Letting $q(t)$ represent the time dependence of the load and substituting for $m(x)$ and $M(x)$ as before results in the equation

$$EIw'''' + \rho A w = C_o\varepsilon_p\frac{\partial^2}{\partial x^2}[H(x - x_1) - H(x - x_2)]q(t) \tag{2.53}$$

$$= C_o\varepsilon_p\frac{\partial^2}{\partial x^2}[\delta'(x - x_1) - \delta'(x - x_2)]q(t). \tag{2.54}$$

To solve this partial differential equation we expand the displacement $w(x,t)$ in an infinite series of the beam's natural modes,

$$w(x,t) = \sum_n W_n \varphi_n(x) q_n(t), \tag{2.55}$$

where the modes $\varphi_n(x)$ are assumed to be known and where the coefficients W_n are to be determined. Substituting this modal series into the equation of motion produces

$$EI \sum_n W_n \varphi_n'''' q_n + \rho A \sum_n W_n \varphi_n q_n = C_o \varepsilon_p [\delta'(x - x_1) - \delta'(x - x_2)] q(t). \tag{2.56}$$

Noting that

$$\varphi_n'''' = \lambda_n^4 \varphi_n \tag{2.57}$$

where the λ_n are known eigenvalues from the related free vibration problem for the beam alone, and assuming the applied voltage is harmonic,

$$q(t) = \hat{q} e^{i\omega t}, \tag{2.58}$$

we simplify the above to

$$EI \sum_n W_n \lambda_n^4 \varphi_n \hat{q}_n - \rho A \omega^2 \sum_n W_n \varphi_n q_n = C_o \varepsilon_p [\delta'(x - x_1) - \delta'(x - x_2)] q(t), \tag{2.59}$$

where now

$$q_n(t) = \hat{q}_n e^{i\omega t}. \tag{2.60}$$

Multiplying eq. (2.59) by another mode $\varphi_m(x)$ and integrating over the length of the beam, and using the orthogonality of the natural modes as expressed by

$$\int_0^L \varphi_m \varphi_n \, dx = \delta_{mn}, \tag{2.61}$$

we can obtain the desired coefficients of the modal series,

$$\hat{W}_n = W_n \hat{q}_n = \frac{-C_0[\varphi'(x_1) + \varphi'(x_2)]}{\rho A \left(\omega_n^2 - \omega^2\right) \left(\int_0^L \varphi_n^2(x)\, dx\right)^2}. \tag{2.62}$$

2.3.3.2 Example: Impulsive Load from a Triangular Patch

As a final example of generating an equivalent transverse load by means of an actuator-induced distributed moment, consider the structural system shown in Fig. 2.16. Attached to the top and bottom surfaces of the beam, and driven by opposite voltages, are identical, triangular PZT elements. These patches are the full width of the beam at its root, $x = 0$, and taper to a point at $x = x_1$.

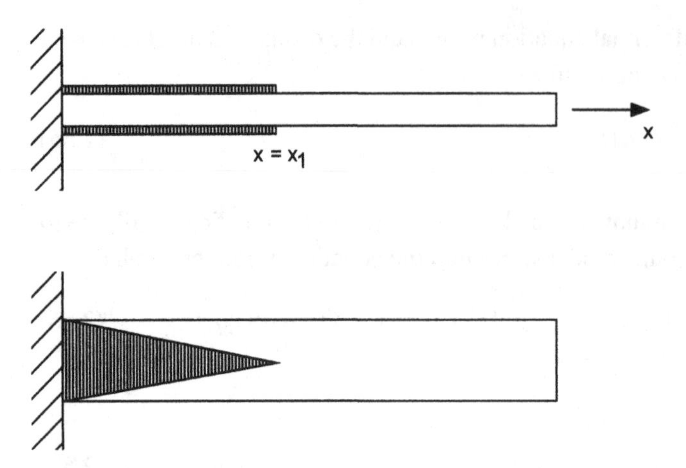

Figure 2.16. Beam driven by triangular PZT actuators.

We assume that the intensity of the moment developed by the actuators is proportional to their width and so varies linearly with x. It is then a simple matter to construct the shear and moment diagram of Fig. 2.17. The bottom curve in that figure shows the derivative of the internal shear force, i.e., external load intensity. (The point moment described in the text, which would otherwise be developed in the beam at the wide end of the actuator patches, is assumed to be reacted by the wall at the root of the cantilevered beam.)

It may be seen that applying a steady voltage to the actuators will produce a static point load on the beam, while applying an electrical impulse to the PZT will cause the beam to be excited by the equivalent of a hammer blow at the point $x = x_1$.

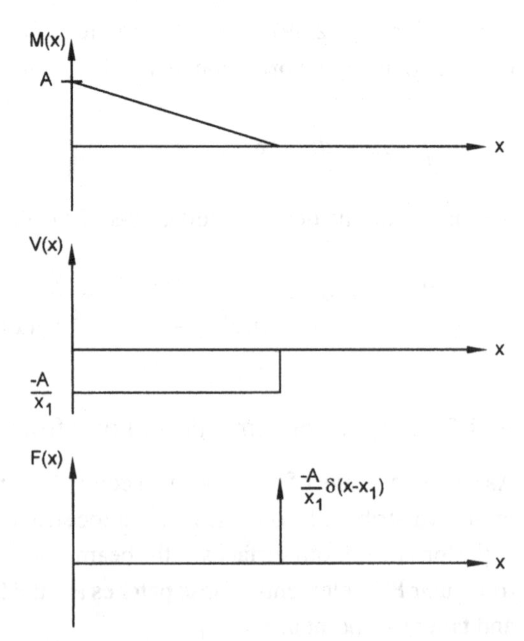

Figure 2.17. Shear and moment diagram for beam with triangular actuator patches.

2.4 Summary

Piezoelectric actuation leading to bending deformation has the following characteristics:

- Better authority can be obtained by attaching crystals on both the top and bottom surfaces of the base structure.
- Perfect bonding is assumed between the crystals and the structure at the interfaces.
- The piezoelectric free strain must exceed in magnitude the maximum strain to be induced during actuation.
- Strain within the crystals is obtained by superposition of the free strain and the strain in the driven structure.
- There is discontinuity of *both* strain and stress at the interfaces.

BIBLIOGRAPHY

Burleigh Instruments, Inc. n.d. *The Piezo Book*, Piezo 330 692 51683–0.
Cady, W. G. 1994. *Piezoelectricity*, Vol. I and II. New York: Dover Publications, Inc.
Crawley, E. F., and E. H. Anderson. 1990. Detailed models of piezoceramic actuation of beams. *Journal of Intelligent Material Systems and Structures* 1(1):4–25.
Holland. 1969. *Design of Resonant Piezoelectric Devices*. Cambridge, Massachusetts: M.I.T. Press.
Cook, W., B. Jaffe, and H. Jaffe. 1971. *Piezoelectric Ceramics*. New York: Academic Press.
Moulson A. J. and J. M. Herbert. 1990. *Electroceramics*. London: Chapman & Hall.
Tierston, H. F. 1969. *Linear Piezoelectric Plate Vibrations*. New York: Plenum Press.

PROBLEMS

1. A specimen of hard PZT has the dimensions and orientation shown in Fig. 2.18.

(a) A voltage $V = 300$ V is applied to the electrodes on the top and bottom of the specimen. Find the resulting in- and out-of-plane strains.

(b) Under the same conditions as in part (a), find the displacement of point P with respect to point O.

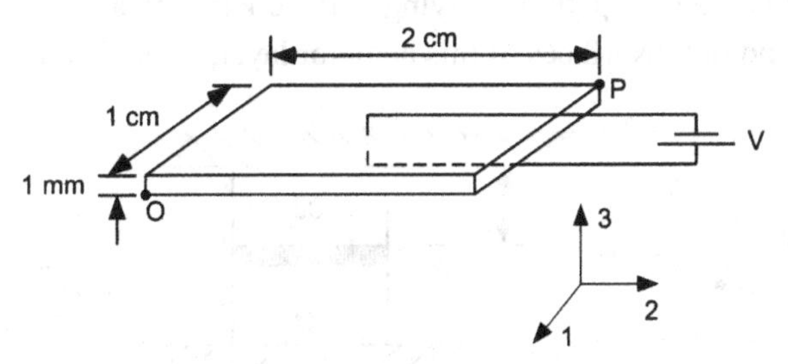

Figure 2.18. PZT specimen for Problem 1.

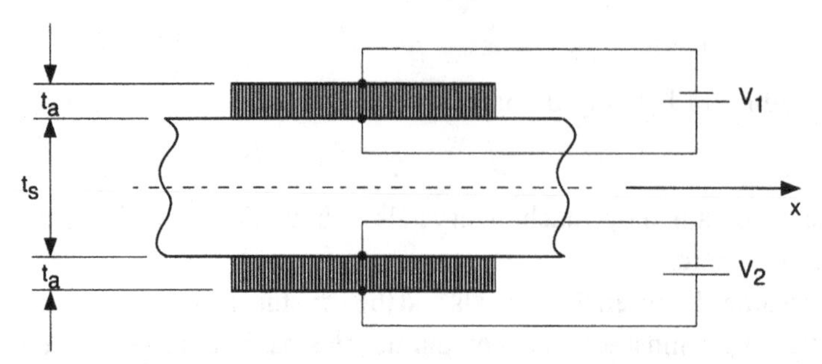

Figure 2.19. One-dimensional structure and PZT crystals of Problem 2.

2. Piezoelectric crystals are attached to the top and bottom surfaces of a one-dimensional structure as shown in Fig. 2.19. The crystals are made of a soft PZT material, and the base structure is made of aluminum. The thicknesses of the base structure and actuators are $t_s = 2\,\text{mm}$ and $t_a = 0.4\,\text{mm}$, respectively. Calculate the stress and strain distributions through the thickness of the structure at the location x_1 and plot the resulting axial or transverse displacement versus x for the conditions below.

(a) $V_1 = V_2 = 500\,\text{V}$

(b) $V_1 = -V_2 = 500\,\text{V}$

3. The PZT specimen of Problem 1 is used as a force transducer to measure loads in the 3-direction.

(a) What will be the voltage produced in response to a uniformly distributed 100 N load?

(b) What load will result in an output of 10 V? What will be the corresponding strains in the crystal?

4. A PZT layer 0.5 mm thick is constrained between blocks of copper, which also act as electrodes. A voltage is applied to the copper blocks so as to produce out-of-plane expansion of the piezoelectric material while the blocks are held between rigid surfaces, as shown in Fig. 2.20. Plot the components of stress and strain in the 3-direction within the PZT for two values of block thickness, $h = 10\,\text{mm}$ and $h = 50\,\text{mm}$.

5. A rod, of radius r_s, is uniformly clad with a layer of PZT of thickness t_a over the portion of its length from x_1 to x_2, and a spatially varying voltage $V(x)$ (steady in time) is applied to the inner and outer surfaces of this actuator layer, as depicted

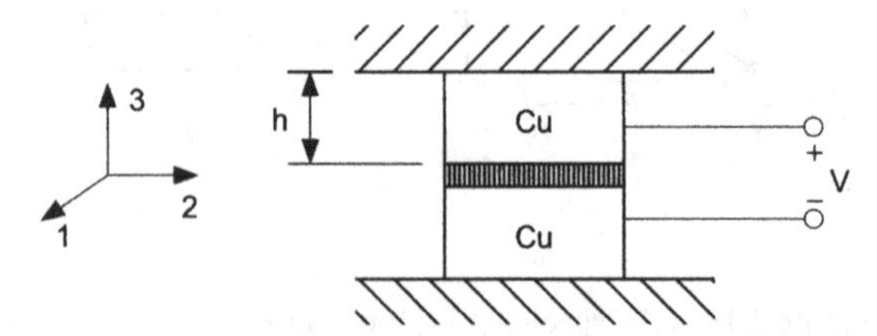

Figure 2.20. PZT layer clamped between copper blocks (Problem 4).

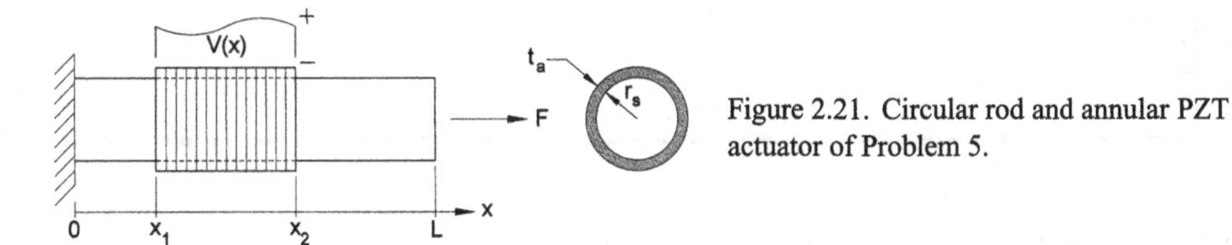

Figure 2.21. Circular rod and annular PZT actuator of Problem 5.

in Fig. 2.21. It is desired to calculate the axial strain in the rod produced by the voltage

$$V(x) = \sin \frac{\pi(x - x_1)}{x_2 - x_1}. \tag{2.63}$$

As before, it is assumed that transverse sections remain plane during deformation.

6. Attached to the outer surface of a circular cylindrical air tank is a hard PZT patch. The tank is made of steel and is 60 cm in diameter with a wall thickness of 4 mm, and the patch is 10 mm square and 0.75 mm thick. If zero gage pressure produces no output from this sensor, what will be the voltage corresponding to an internal pressure 1 000 kPa above atmospheric?

7. A pair of PZT patches is to be designed to excite bending vibration of a beam by the mechanism discussed in Section 2.3.3. The beam is 80 cm long and 3 × 20 mm in cross section, and is made of aluminum. The actuators should be capable of producing a peak-to-peak motion of 1 mm at the anti-node of the lowest mode of the structure when driven by an electric field of ±100 000 V/m, and should be no thicker than 1 mm. Determine whether the patches should be hard or soft PZT and what portion of the beam they should be bonded to. Will the configuration chosen be effective in exciting the second bending mode of the beam?

3

Shape Memory Alloys

3.1 Introduction

Shape memory alloys possess an interesting property by which the metal "remembers" its original size or shape and reverts to it at a characteristic transformation temperature. This feature, known as shape memory effect, was first observed in samples of gold-cadmium in 1932 and 1951, and in brass (copper-zinc) in 1938. It was not until 1962, however, that William J. Buehler and coworkers at the Naval Ordnance Laboratory (NOL) discovered that nickel-titanium showed this shape memory effect. They found that an alloy consisting of equal numbers of nickel and titanium atoms showed the transformation that leads to shape memory. Further, by adding a slight extra amount of nickel in the alloy they could change the transformation temperature from near 100 °C down to below 0 °C. Also, this alloy had constituents that were not prohibitively expensive, had greater shape memory strain (up to 8%) than other alloys, and could be fabricated with existing metalworking techniques.

These alloys change phase at certain critical temperatures and therefore they display different stress-strain characteristics in different temperature ranges. Further, even plastic deformations induced at a lower temperature can be completely recovered at a higher temperature. Before we begin to probe further to understand the properties of such materials, it is interesting to note the story behind the discovery of this phenomenon.

In a recent article, Kauffman and Mayo (1993) describe the research activities of Buehler at the Naval Ordnance Laboratory in 1956 in the following manner. As NOL's supervisory physical metallurgist, Buehler was engaged in testing intermetallic compounds for the nose cone of the Navy's missile SUBROC. The goal was to find a metal with a high melting point and high impact resistant properties. From among sixty compounds, Buehler selected twelve candidates to measure their impact resistance by hitting arc-melted buttons with a hammer. From these tests, he noted that a nickel-titanium alloy seemed to exhibit the greatest resistance to impact in addition to satisfactory properties of elasticity, malleability, and fatigue. He named it NiTiNOL to include the acronym of the name of his laboratory.

While studying the properties of this material with varying percentages of nickel and titanium, Buehler made an observation that was at the root of the discovery of an extraordinary property. One day in 1959, Buehler and his assistant cast six NiTiNOL bars in their arc-melting furnace and laid them out on a table to cool. He took one of them, the one that came out first and had cooled off, to a shop grinder to smooth the surface. On the way, he intentionally dropped it on the concrete floor out of curiosity and noticed it made a dull thud indicative of high damping. Puzzled, he tested the other bars and found that they produced a bell-like quality sound. "Following this I literally ran with one of the warmer bars to the closest source of cold water – the drinking fountain – and chilled the warm bar. After thorough cooling, the bar was once again dropped on the floor. To my continued amazement it now exhibited the leaden-like acoustic response. To confirm this unique change, the cooled bars were heated through in boiling water, and they now rang brilliantly when dropped upon the concrete floor." Buehler knew that the acoustic damping signaled a change in the atomic structure that can be turned off and on by simple heating and cooling near room temperature. He did not yet know that this rearrangement in the atomic structure would later result in another, even more interesting, phenomenon.

It was in 1960 that he was joined by Raymond Wiley, from an NOL group that worked on failure analysis of various metals. Wiley would generate much of the data essential for understanding NiTiNOL's properties. Wiley demonstrated to his management the fatigue resistance by repeatedly flexing a NiTiNOL wire, a hundredth inch thick, which had been bent into short accordion-like folds. The directors who were present at this meeting passed the strip around the table, repeatedly flexing and unflexing it and were impressed with how well it held up. It was at this meeting serendipity played a major role. One of the associated technical directors, David Muzzey, "decided to see how it would behave under heat. Muzzey was a pipe smoker, so he held the compressed NiTiNOL strip in the flame of his lighter. To the great amazement of all, it stretched out longitudinally. When Buehler heard about the incident, he realized that it had to be related to the acoustic behavior he had noted earlier. While the sonic phenomenon was essentially an interesting curiosity, it was clear from the start that shape memory could have very important applications" (Kauffman and Mayo, 1993).

Thus, NiTiNOL has a low temperature phase and a high temperature phase and its unique properties arise from a change in its phase. The phase change is between two solid phases and involves rearrangement of atoms within the crystal lattice (Fig. 3.1). The internal structure is different at different temperatures. The low temperature phase is known as martensite (named after the German metallographer Adolf Martens), with a highly twinned crystalline structure, and the high temperature phase is called austenite (named after William Chandler Austen, an English metallurgist), with a body-centered cubic structure. The critical temperatures at which the phase transformations take place are identified as M_s, M_f, A_s, and A_f, which represent the temperatures at the start of martensite, finish of martensite, start of austenite, and finish of austenite transformation. Recovery begins at A_s and is completed at A_f. Any method of heating is adequate. There is no stress involved in the process described above and the deformation mode

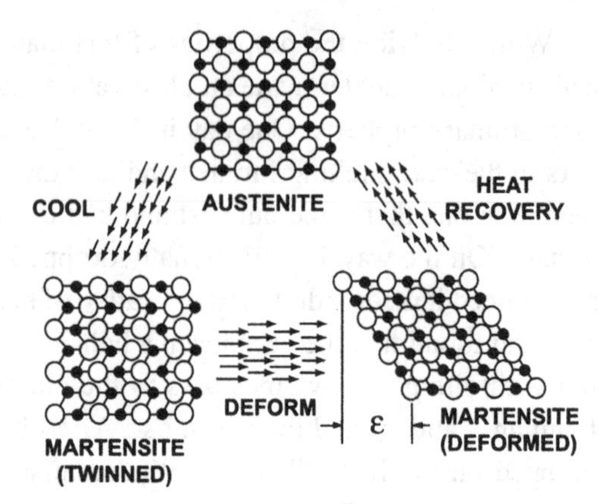

Figure 3.1. Shift in crystal structure accompanying phase change in shape memory alloys. (Courtesy of Raychem Corporation.)

is nondamaging to the crystal structure because of the self-accommodating nature of deformation. The memory discussed above is one way (Fig. 3.2). With two-way memory (Fig. 3.3), the metal "remembers" two shapes, each of which can be recovered at a different temperature. The illustrations of Figs. 3.4 through 3.6 help clarify these concepts.

Up to 7% strain can generally be reversed, and in some cases as much as 10%. Typical properties of some shape memory alloys (SMAs) are listed in Tables 3.1 through 3.4.

3.2 Experimental Phenomenology

Let us say that we are able to put an SMA wire through a temperature cycle as shown in Fig. 3.7. Start with the wire in a fully austenitic phase, that is, the "high temperature" phase. Let A_f represent this temperature. Now cool the wire until it reaches a temperature at which a phase transformation begins. This transformation is, of course, from a fully austenitic phase to martensitic phase. Let M_s represent this temperature to indicate the starting of the martensitic phase, i.e., the "low temperature" phase. At M_s, the material begins to change its phase to martensite. Upon further cooling, the martensite plates begin to increase until M_f, at which temperature the wire is in a fully martensitic state. If ξ represents the fraction of martensite in the material, then at A_f and M_s, $\xi = 0$, and $\xi = 1$ at M_f and A_s. If we now begin to heat the wire when

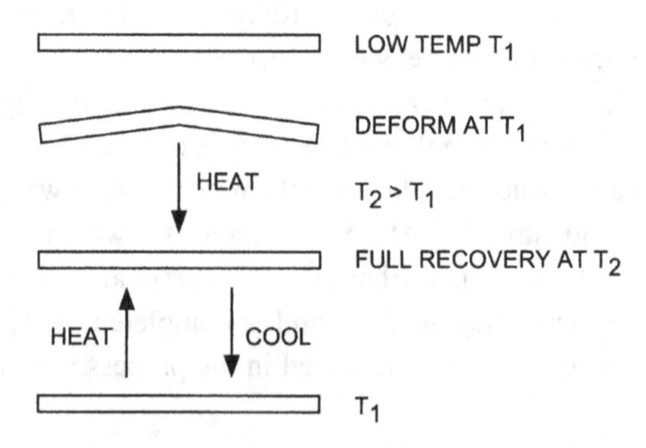

Figure 3.2. One-way shape memory (after Wayman, 1993).

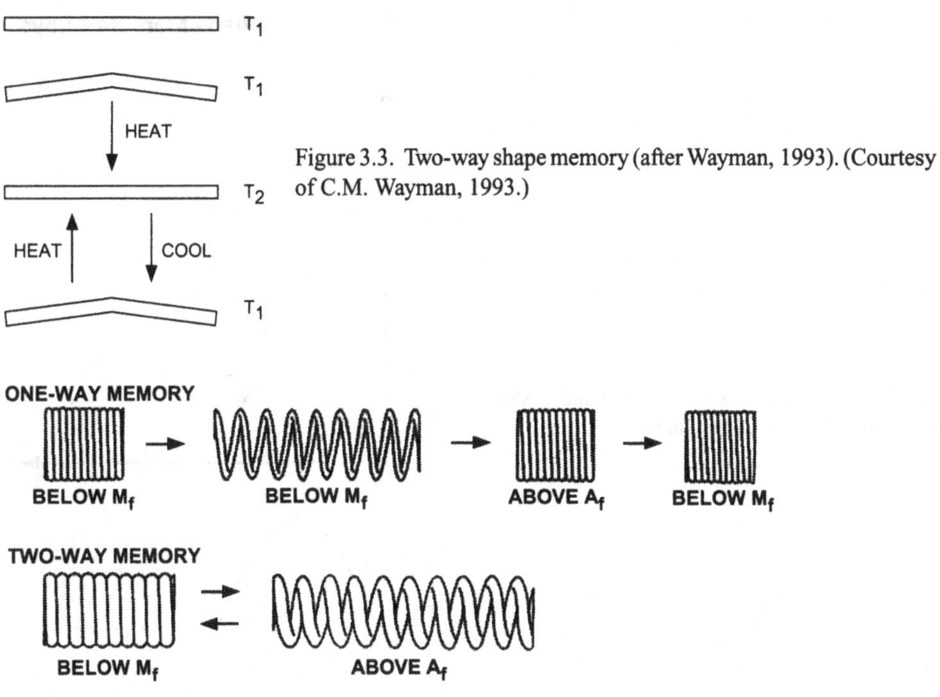

Figure 3.3. Two-way shape memory (after Wayman, 1993). (Courtesy of C.M. Wayman, 1993.)

Figure 3.4. Examples of one-way and two-way memory. (Reprinted, by permission, from Materials Research Society. Courtesy of C. M. Wayman, 1993.)

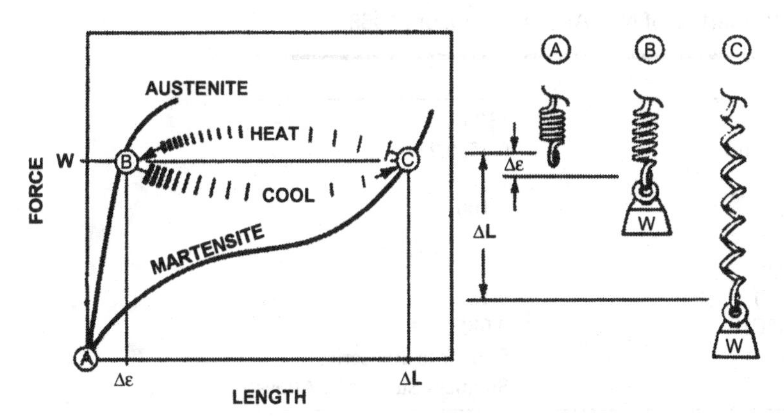

Figure 3.5. SMA structural responses at two different phases of the material. (Courtesy of Raychem Corporation.)

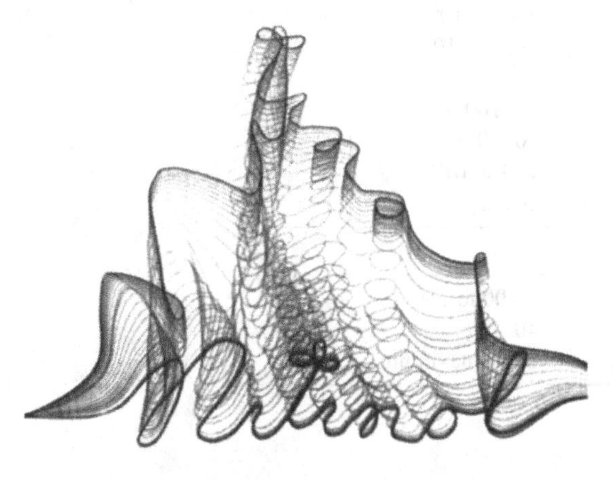

Figure 3.6. A crumpled wire expands into the word "Nitinol" on heating (Kauffman and Mayo, 1993). (U.S. Army photograph.)

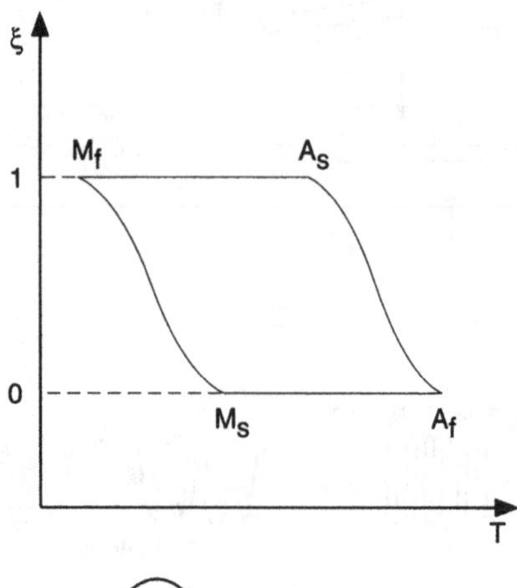

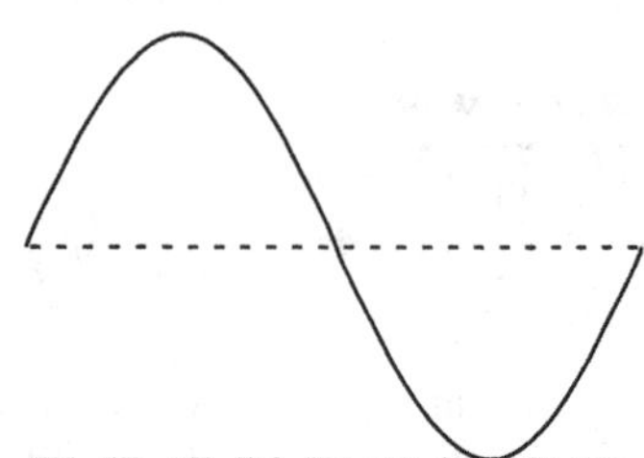

Figure 3.7. Mathematical modeling of martensitic fraction in an SMA wire at critical temperatures.

Table 3.1. Typical Properties of NiTi Alloys (Hodgson, 1988)

Physical Properties	
Melting Point (°C)	1300
Density (gm/cc; lb/cu. in.)	6.45; 0.233
Electrical Resistivity:	
Austenite ($\mu\Omega$ cm)	~100
Martensite ($\mu\Omega$ cm)	~70
Thermal Conductivity:	
Austenite (W/cm °C)	0.18
Martensite (W/cm °C)	0.085
Corrosion Resistance	Similar to 300 series Stainless Steel or Ti Alloys
Mechanical Properties	
Young's Modulus:	
Austenite (psi)	~12×10^6
Martensite (psi)	~4.6×10^6
Yield Strength:	
Austenite (psi)	$28–100 \times 10^3$
Martensite (psi)	$10–20 \times 10^3$
Ultimate Strength (psi)	130×10^3
Elongation at Failure (%)	20-30
Transformation Properties	
Transformation Temperature (°C)	–200 to 110
Latent Heat of Transformation (cal/g atom)	40
Shape memory Strain	8.4% maximum

Table 3.2. Properties of Shape Memory Alloys. (Courtesy of Memry Technologies Inc.)

Physical Properties	**Cu-Zn-Al**	**Cu-Al-Ni**	**Ni-Ti**
Density (g/m³)	6.45	7.64	6.4
Resistivity ($\mu\Omega$ cm)	8.5–9.7	11–13	80–100
Thermal Conductivity (J/m s K)	120	30–43	10
Heat Capacity (J/kg K)	400	373–574	390
Mechanical Properties	**Cu-Zn**	**Cu-Al-Ni**	**Ni-Ti**
Young's Modulus (GPa):			
β-phase	72	85	83
Martensite	70	80	34
Yield Strength (MPa):			
β-phase	350	400	690
Martensite	80	130	70–150
Ultimate Tensile Strength (MPa)	600	500–800	900
Transformation Properties	**Cu-Zn-Al**	**Cu-Al-Ni**	**Ni-Ti**
Heat of Transformation (J/mole):			
Martensite	160–440	310–470	295
R-phase	—	—	55
Hysteresis (K):			
Martensite	10–25	15–20	30–40
R-phase	—	—	2–5
Recoverable Strain (%):			
One-Way (Martensite)	4	4	8
One-Way (R-phase)	—	—	0.5–1
Two-Way (Martensite)	2	2	3

$\xi = 1$, nothing happens to the material phase, but the temperature of the wire increases. When the temperature reaches A_s, martensite plates begin to rearrange themselves into their original configuration. Thus, A_s represents the temperature at which the "high temperature" phase starts. Upon further increase in temperature, the rearrangement continues and is complete at A_f, when the wire is in a fully austenitic state. The cycle is thus complete with the wire beginning in the austenitic phase, transforming at a "lower temperature" to fully martensitic phase, and reverting, upon heating, to the original "high temperature" or parent phase. Thus, the critical temperatures are M_s, M_f, A_s, A_f, with the martensite fraction changing from $\xi = 0$ to $\xi = 1$. Note that throughout the transformation discussed above, no stress was applied. The cycle is entirely driven by temperature. The influence of stress on this cycle will be discussed later. The above observations can be represented graphically as shown in Fig. 3.7 and can be used to obtain a mathematical representation of this behavior.

Let us examine the graphical representation and find a mathematical expression that describes this behavior. Note how a cosine function may describe the observation. However, the abscissa should be pushed down to the —— line from the ---- line in order to reach a value of $\xi = 0$ at $T = \pi$. Thus, we wish to freeze the temperature scale but fit a cosine such that the function has values $\xi = 1$ at $\theta = 0$ and $\xi = 0$ at $\theta = \pi$.

Table 3.3. Properties of Shape Memory Alloys (Cronauer and Perkins, 1987)

Metal	Specific Damping Capacity (%)	Yield Strength (10^3 psi)	Density (gm/cm^3)
Magnesium (wrought)	49	26	1.74
Cu-Mn alloys (Incramute, Sonoston)	40	45	7.5
Ni-Ti alloy (NiTiNOL)	40	25	6.45
Fe-Cr-Al alloy (Silentalloy)	40	40	7.4
High-C gray iron	19	25	7.7
Nickel (pure)	18	9	8.9
Iron (pure)	16	10	7.86
Martensitic stainless steel	8	85	7.7
Gray cast iron	6	25	7.8
SAP (aluminum powder)	5	20	2.55
Low-carbon steel	4	50	7.86
Ferritic stainless steel	3	45	7.75
Malleable, nodular cast irons	2	50	7.8
Medium-carbon steel	1	60	7.86
Austenitic stainless steel	1	35	7.8
1100 Aluminum	0.3	5	2.71
Aluminum alloy 2024–T4	<0.2	47	2.77
Nickel-base superalloys	<0.2	Range	8.5
Titanium alloys	<0.2	Range	4.5
Brasses, bronzes	<0.2	Range	8.5

Let

$$\xi = C_1 \cos \alpha_M (T - M_f) + C_2 = C_1 \cos \theta + C_2. \tag{3.1}$$

This representation and analysis follow closely those of Liang and Rogers (1990). Then

$$\xi = 0 \quad \text{at} \quad \theta = \pi \Rightarrow C_1 = C_2 \tag{3.2}$$

$$\xi = 1 \quad \text{at} \quad \theta = 0 \Rightarrow C_1 = C_2 = \frac{1}{2} \tag{3.3}$$

and, thus, the transformation from austenite to martensite ($A \rightarrow M$) may be represented as

$$\xi = \frac{1}{2} \cos \alpha_M (T - M_f) + \frac{1}{2}. \tag{3.4}$$

As $\xi = 0$ at $\theta = \alpha_M (T - M_f) = \pi$,

$$\alpha_M = \frac{\pi}{M_s - M_f}, \qquad M_f \le T \le M_s. \tag{3.5}$$

Similarly, the other end of the graphical representation ($M \rightarrow A$) can be written as

$$\xi = \frac{1}{2} \cos \alpha_A (T - A_s) + \frac{1}{2}, \tag{3.6}$$

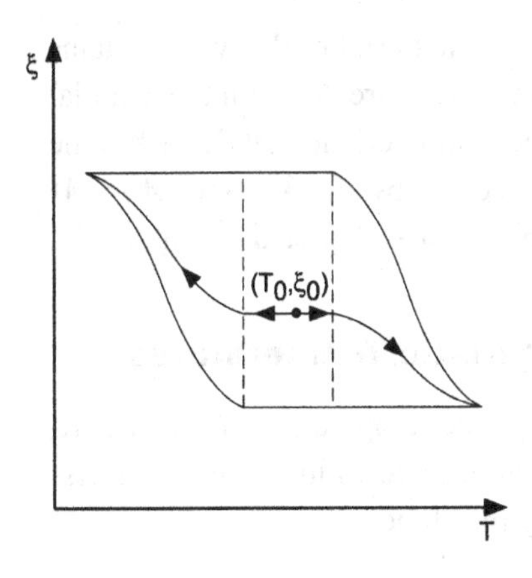

Figure 3.8. Martensitic fraction in SMA for prescribed initial conditions.

where

$$\alpha_A = \frac{\pi}{A_f - A_s}, \quad A_s \leq T \leq A_f. \tag{3.7}$$

The above transformations and their mathematical representations were *on* the characteristic curves for the transformation, as a result of starting from fully austenitic or fully martensitic initial conditions. However, if we were to begin the process at some value ξ_0 at a corresponding T_0, then the above expressions can be modified as follows. The equations are obtained on the basis that no new martensite is added until the temperature T reaches M_s (see Fig. 3.8).

Application of the boundary conditions $\xi = 0$ at $\theta = 0$ and $\xi = \xi_0$ at $\theta = \pi$ to eq. (3.1) leads to the constants

$$C_1 = \frac{1 - \xi_0}{2}, \tag{3.8}$$

$$C_2 = \frac{1 + \xi_0}{2}. \tag{3.9}$$

Therefore,

$$\xi = \frac{1 - \xi_0}{2} \cos[\alpha_M(T - M_f)] + \frac{1 + \xi_0}{2} \quad \text{for} \quad A \rightarrow M. \tag{3.10}$$

For transformation from $M \rightarrow A$ beginning with $\xi = \xi_0$ and $T = T_0$ on the basis of the assumption that no new austenite is added until the temperature reaches A_s, we can write

$$\xi = 0 \quad \text{at} \quad \alpha = \pi \Rightarrow C_1 = C_2. \tag{3.11}$$

However,

$$\xi = \xi_0 \quad \text{at} \quad \theta = 0 \Rightarrow C_1 = \frac{\xi_0}{2} = C_2, \tag{3.12}$$

and therefore,

$$\xi = \frac{\xi_0}{2}[\cos \alpha_A(T - A_s) + 1]. \tag{3.13}$$

These characteristics are shown graphically in Fig. 3.8.

Initial conditions ξ_0, T_0 represent a condition that the material of the wire contains some martensite, ξ_0, and some austenite, $1 - \xi_0$, at a temperature T_0. With these initial conditions, if the material were to be cooled, then the characteristic that describes the change in the state variable ξ can be shown to be represented by eq. (3.10) for $A \to M$ and by eq. (3.13) for $M \to A$ transformation. This behavior is depicted by Fig. 3.8.

3.3 Influence of Stress on the Characteristic Temperatures

Experimental observations indicate that the characteristic temperatures M_f, M_s, A_s, and A_f (i.e., the temperatures at which the phase changes occur) increase with stress as shown in Fig. 3.9. These changes are described by the slopes

$$C_M = \tan\alpha, \tag{3.14}$$

$$C_A = \tan\beta. \tag{3.15}$$

Generally, it is assumed that $\alpha = \beta$ and M_f, M_s, A_s, A_f are the critical temperatures at which phase changes occur at $\sigma = 0$. With $\sigma \neq 0$, higher temperatures will be needed to bring about a phase change. The increase is linear, with critical temperatures increasing with applied stress. For $M \to A$ transformation with $\sigma \neq 0$,

$$\xi = \frac{\xi_0}{2}\{\cos[\alpha_A(T - A_s) + b_A\sigma] + 1\}. \tag{3.16}$$

This equation reflects an increased temperature proportional to σ. With reference to Fig. 3.9 we can calculate b_A as follows: with $\xi = 0$ at $\sigma \neq 0$ and $T = A_f^*$,

$$\alpha_A(A_f^* - A_s) + b_A\sigma = \pi, \tag{3.17}$$

$$\alpha_A(A_f - A_s + A_{f_\sigma}) + b_A \cdot \sigma = \pi. \tag{3.18}$$

Recall

$$\alpha_A = \frac{\pi}{A_f - A_s}; \qquad \alpha_A\left(\frac{\pi}{\alpha_A} + A_{f_\sigma}\right) + b_A\sigma = \pi; \tag{3.19}$$

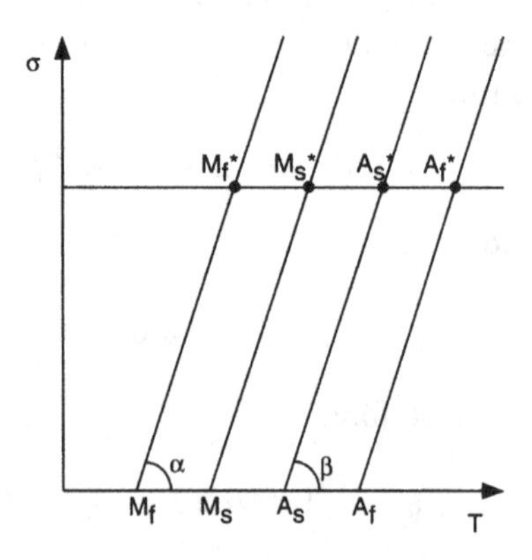

Figure 3.9. Influence of stress on critical phase change temperatures.

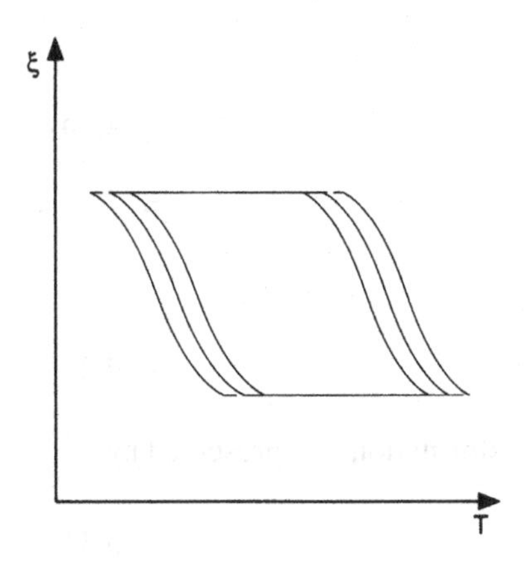

Figure 3.10. Shift of ξ-T characteristic to the right with increased stress.

thus

$$b_A = \frac{-\alpha_A}{C_A}. \tag{3.20}$$

Similarly,

$$b_M = \frac{\alpha_M}{C_M}. \tag{3.21}$$

Thus, the effect of increased stress is to shift the characteristic to the right in Fig. 3.10 with

$$\xi = \frac{1-\xi_0}{2}\cos[\alpha_M(T - M_f) + b_M\sigma] + \frac{1+\xi_0}{2} \tag{3.22}$$

for $A \to M$ transformation. For $M \to A$ transformation, from eq. (3.16),

$$\xi = \frac{\xi_0}{2}\{1 + \cos\theta\}; \quad \theta = \alpha_A(T - A_s) + b_A\sigma; \quad 0 \le \theta \le \pi. \tag{3.23}$$

The range for σ to produce stress-induced martensite (SIM) can be found to be

$$0 \le \alpha_A(T - A_s) + b_A\sigma \le \pi. \tag{3.24}$$

Thus,

$$\sigma \le C_A(T - A_s) \tag{3.25}$$

represents the upper limit for σ.

The lower limit can similarly be found with

$$\alpha_A(T - A_s) + b_A\sigma = \pi, \tag{3.26}$$

$$C_A(T - A_s) + \frac{\pi}{b_A} = \sigma. \tag{3.27}$$

Therefore,

$$C_A(T - A_S) - \frac{\pi}{|b_A|} \le \sigma \tag{3.28}$$

and thus

$$C_A(T - A_s) - \frac{\pi}{|b_A|} \leq \sigma \leq C_A(T - A_s) \tag{3.29}$$

holds for the $M \to A$ transformation.

Similarly,

$$C_M(T - M_f) - \frac{\pi}{|b_M|} \leq \sigma \leq C_M(T - M_f) \tag{3.30}$$

for the $A \to M$ transformation. With $\sigma \neq 0$, this transformation is represented by

$$\xi = \frac{\xi_0}{2}\{\cos[\alpha_A(T - A_s) + b_A\sigma] + 1\}, \tag{3.31}$$

which reduces to the simpler equation shown earlier when $\sigma = 0$.

The cosine function for ξ allows us to determine the range of stress within which these transformations may take place. With $b_A = -\alpha_A/C_A$ and $b_M = -\alpha_M/C_M$, the range can be shown to be, for $A \to M$,

$$C_M(T - M_f) - \frac{\pi}{|b_M|} \leq \sigma \leq C_M(T - M_f), \tag{3.32}$$

and for the reverse transformation $M \to A$,

$$C_A(T - A_s) - \frac{\pi}{|b_A|} \leq \sigma \leq C_A(T - A_s). \tag{3.33}$$

These define the ranges in which stress-induced transformations may take place.

Consider

$$C_M(T - M_f) - \frac{\pi}{|b_M|} \leq \sigma. \tag{3.34}$$

Then

$$\sigma = C_M(T - M_f) - \frac{\pi}{|b_M|} \tag{3.35}$$

represents the lower limit for σ above which a fully austenitic wire will begin to experience martensitic phase. Let us define this lower limit as $\sigma = \bar{\sigma}_{\text{lin}}$. Thus,

$$\bar{\sigma}_{\text{lin}} = C_M(T - M_f) - \frac{\pi}{|b_M|}, \tag{3.36}$$

and because

$$b_M = -\frac{\alpha_M}{C_M}, \tag{3.37}$$

$$\bar{\sigma}_{\text{lin}} = C_M(T - M_f) - \frac{\pi C_M}{\alpha_M}, \tag{3.38}$$

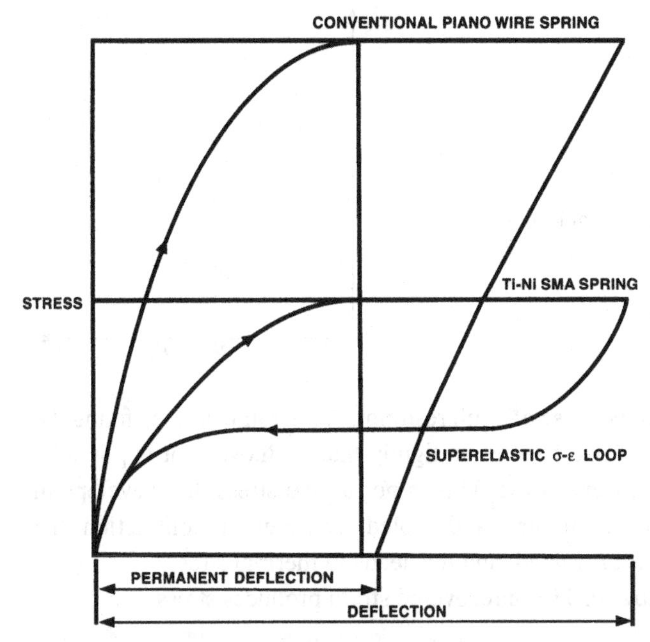

Figure 3.11. Superelastic stress-strain response of shape memory alloys. (Reprinted, by permission, from Materials Research Society. Courtesy of C. M. Wayman, 1993.)

However, $\alpha_M = \frac{\pi}{M_s - M_f}$, so

$$\bar{\sigma}_{\text{lin}} = C_M(T - M_f) - \frac{\pi C_M}{\pi/(M_s - M_f)} \tag{3.39}$$

$$= C_M(T - M_f) - C_M(M_s - M_f) \tag{3.40}$$

$$= C_M(T - M_s). \tag{3.41}$$

The subscript lin also represents the limit of linearity for the stress above which the stress-strain relationship will be nonlinear because of nonzero martensite.

$$\bar{\varepsilon}_{\text{lin}} = \text{corresponding limit for linear strain} = \frac{\bar{\sigma}_{\text{lin}}}{E}. \tag{3.42}$$

Thus, phase transformations resulting entirely from stress and leading to hysteresis loops of the type shown in Fig. 3.11 are another important characteristic of the material, referred to as superelasticity.

Whenever a wire made of SMA is stretched at a low temperature, some level of martensite may remain upon loading if the load satisfies the range requirements defined above. Upon heating the wire, the martensitic residual strain may be recovered. The extent of this recovery will depend upon the extent of restraint, if any. There are three distinct possibilities.

1. Full recovery can occur with no external load on the wire. No work is done in the process.

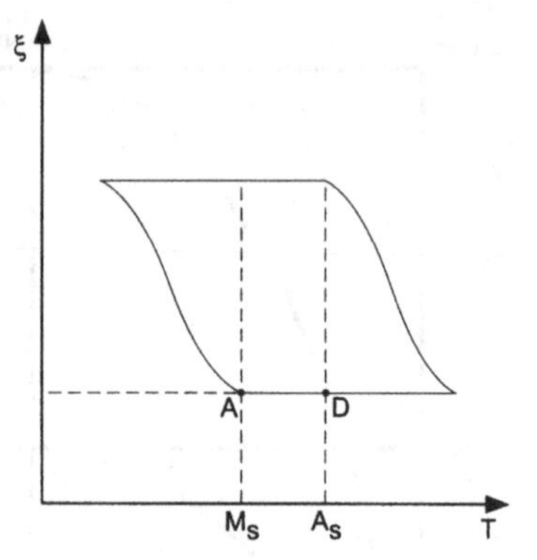

Figure 3.12. Isothermal mechanical loading of a wire.

2. No recovery is allowed because of a full restraint. A restraint, such as fixing the elongated wire between two fixtures so tightly that no movement is possible, will generate large recovery stress. This is not unlike stress that develops in a steel wire fixed between fixtures and cooled. The thermal contraction that would have occurred is prevented, and this leads to thermal stress.
3. Partial recovery is allowed. The unrecovered strain produces stress.

Thus, a shape memory alloy wire behaves differently depending upon the level of restraint that controls its recovery of martensitic strains.

3.4 Constitutive Modeling of the Shape Memory Effect

Consider an SMA wire subjected to an isothermal mechanical loading and unloading. Assume the wire is in a 100% austenitic state and at a temperature between M_s and A_s, that is, along the line DA in Fig. 3.12. Before representing the constitutive behavior, observe that the stress state in an SMA component is a function of three primary state variables. They are ξ, the fraction of martensite; T, the temperature at which the component is operating; and ε, the strain at which the component is functioning. Therefore, $\bar{\sigma} = \bar{\sigma}(\bar{\varepsilon}, T, \xi)$. Note that $\bar{\sigma}$ is the Piola-Kirchoff stress and $\bar{\varepsilon}$ is Green strain.

$$\dot{\bar{\sigma}} = \frac{\partial \bar{\sigma}}{\partial \bar{\varepsilon}}\dot{\bar{\varepsilon}} + \frac{\partial \bar{\sigma}}{\partial T}\dot{T} + \frac{\partial \bar{\sigma}}{\partial \xi}\dot{\xi} \tag{3.43}$$

Integrating with respect to time from the initial conditions $\bar{\varepsilon}_0$, T_0, ξ_0, we can write a unified constitutive relation

$$\bar{\sigma} - \bar{\sigma}_0 = D(\bar{\varepsilon} - \bar{\varepsilon}_0) + \theta(T - T_0) + \Omega(\xi - \xi_0), \tag{3.44}$$

where D = Young's modulus, θ = thermoelastic tensor, and Ω = transformation tensor.

An isothermal condition implies $T = T_0$, $\sigma < C_M(T - M_s)$, $\xi = \xi_0$ (i.e., not fully austenitic). Assume zero initial strain, $\bar{\varepsilon}_0 = 0$, at $t = 0$, $\bar{\sigma}_0 = 0$, $\bar{\xi}_0 = 0$, $T = T_0$. Under these conditions, if we load and unload, we are going to be in the linear region in which $\bar{\sigma} = D\bar{\varepsilon}$ and the linear elastic limit is

$$\bar{\sigma}_{\text{lin}} = C_M(T - M_s) \tag{3.45}$$

and

$$\bar{\varepsilon}_{\text{lin}} = \frac{\bar{\sigma}_{\text{lin}}}{D}. \tag{3.46}$$

When $\bar{\sigma} > \bar{\sigma}_0$ (i.e., the applied stress exceeds $\bar{\sigma}_{\text{lin}}$), the excess stress induces martensite and the constitutive relation becomes

$$\bar{\sigma} - \bar{\sigma}_0 = D(\bar{\varepsilon} - \bar{\varepsilon}_0) + \Omega(\bar{\xi} - \bar{\xi}_0), \tag{3.47}$$

with $\bar{\sigma}_0 = \bar{\sigma}_{\text{lin}}$, $\bar{\varepsilon}_0 = \bar{\varepsilon}_{\text{lin}}$, $\bar{\xi} = 0$ and

$$\bar{\sigma} - \bar{\sigma}_{\text{lin}} = D\bar{\varepsilon} - D\bar{\varepsilon}_{\text{lin}} + \Omega\bar{\xi}. \tag{3.48}$$

Because

$$\bar{\sigma} = D\bar{\varepsilon} + \Omega\bar{\xi}, \tag{3.49}$$

$$\bar{\sigma}_{\text{lin}} = D\bar{\varepsilon}_{\text{lin}}. \tag{3.50}$$

Thus, the governing equations when $\sigma > \sigma_{\text{lin}}$ are

$$\bar{\sigma} = D\bar{\varepsilon} + \Omega\bar{\xi}, \tag{3.51}$$

$$\bar{\sigma}_{\text{lin}} = C_m(T - M_s), \tag{3.52}$$

$$\bar{\varepsilon}_{\text{lin}} = \frac{\bar{\sigma}_{\text{lin}}}{D}, \tag{3.53}$$

and

$$\xi = \frac{1 - \xi_A}{2}\cos[\alpha_m(T - M_f) + b_M\sigma] + \frac{1 + \xi_A}{2} \tag{3.54}$$

for the $A \rightarrow M$ transformation.

How can we determine Ω? Because we consider loading and unloading, let us unload the wire when $\xi = 1$. Recall the ambient temperature is less than A_s. After the unloading, the stress reaches zero, but the strain remains at the maximum limit it attained under σ_{lin}. It is a linear martensitic elastic unloading, as shown in Fig. 3.13.

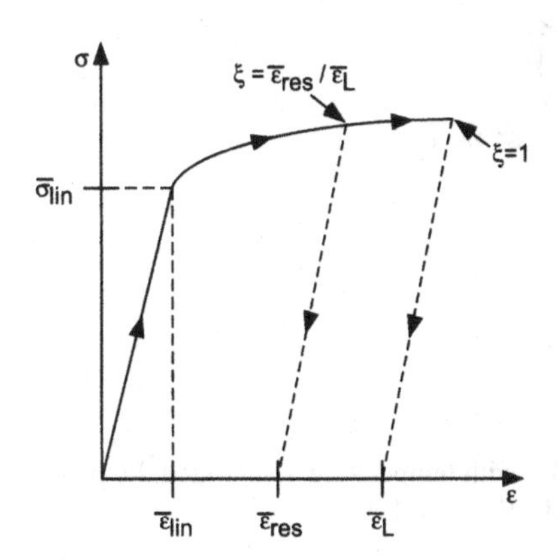

Figure 3.13. Stress-strain for unloading a wire (after Liang and Rogers, 1990).

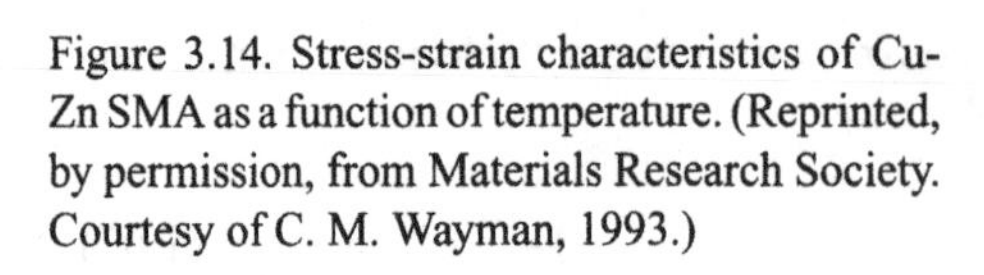

Figure 3.14. Stress-strain characteristics of Cu-Zn SMA as a function of temperature. (Reprinted, by permission, from Materials Research Society. Courtesy of C. M. Wayman, 1993.)

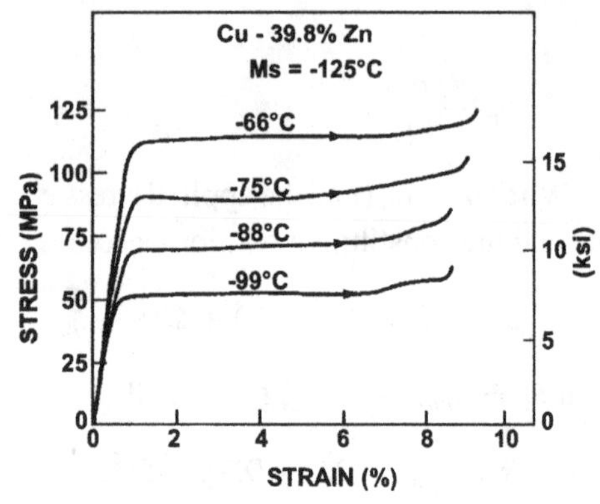

With $\xi = 1$,

$$\bar{\sigma} = 0 = D\bar{\varepsilon}_{\text{lin}} + \Omega, \tag{3.55}$$

$$\bar{\varepsilon}_L = -\frac{\Omega}{D}, \tag{3.56}$$

and

$$\bar{\varepsilon}_{\text{res}} = -\frac{\Omega}{D}\xi = \bar{\varepsilon}_L. \tag{3.57}$$

Thus

$$\xi = -\bar{\varepsilon}_{\text{res}}\frac{D}{\Omega} = -\bar{\varepsilon}_{\text{res}}\left(\frac{-1}{\bar{\varepsilon}_L}\right) = \frac{\bar{\varepsilon}_{\text{res}}}{\bar{\varepsilon}_L}, \tag{3.58}$$

where ξ is SIM, the stress-induced martensite.

Typical stress-strain characteristics of shape memory alloys at various temperatures are illustrated in Fig. 3.14. A comprehensive view of the behavior of shape memory alloys in the stress-strain-temperature domain is provided in Fig. 3.15 and is useful to appreciate the basic characteristics of this material.

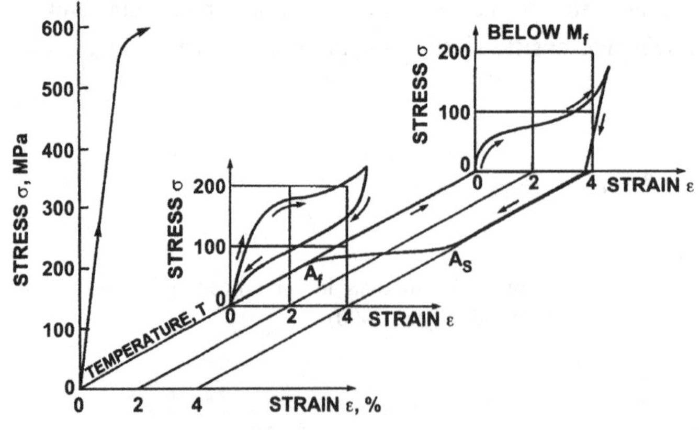

Figure 3.15. Variation of stress-strain characteristics of SMA with temperature. (Reprinted, by permission, from Materials Research Society. Courtesy of C. M. Wayman, 1993.)

3.4.1 Design Considerations: A Simple Example

The example below is instructive in the type of considerations that are important in designing structural components in which shape memory alloys can be used as actuators. The illustrative problem is to design an extension spring needed to lift a load of 1 lb through a distance of 2 in. in 5 s. We shall use the properties of Tinel listed in Table 3.4.

If it were a straight wire, then $P_{max} = (60\,\text{ksi})(\pi/4)d^2 = 47{,}000d^2$, where d is the diameter of the spring wire and $\delta L = \delta\varepsilon = 0.08L$ for an 8% strain. However, we need to design a helical spring, for which some design charts (courtesy Raychem) are helpful. Figure 3.16 can be used to obtain d if the spring index is chosen. If we choose D/d to be 5, where D = diameter of spring, then $d = 0.3$ in. to carry a load of 1 lb. With Fig. 3.17, we can calculate the number of turns in the helical spring because $(\delta L/N) = (0.24)(0.03)(25)$ and $\delta L = 2$ in. Thus, the desired spring has twelve turns of wire of diameter 0.03 in. and $D/d = 5$.

The mechanical design is complete. But the actuator design is incomplete until we calculate the power required and temperatures induced to ensure they are within

Table 3.4. Nominal Tinel Properties. (Courtesy of Raychem Corporation.)

Mechanical (Austenite)	
Young's modulus (tension)	14×10^6 psi
Yield strength (tension)	60×10^3 psi
Ultimate strength (tension)	125×10^3 psi
Poisson's ratio	0.33
Elongation at failure	15%–20%
Maximum sustained temperature	600 °F
Shape Memory (No Load)	
A_s	175 °F
A_f	200 °F
M_s	160 °F
M_f	140 °F
Available shape memory	8%
Physical	
Density	0.235 lb/in^3
Thermal conductivity	10.4 BTU/hr ft °F
Specific heat	0.20 BTU/lb °F
Heat of transformation	10.4 BTU/lb
Coefficient of thermal expansion	
Martensite	3.67×10^{-6}/°F
Austenite	6.11×10^{-6}/°F
Electrical	
Resistivity	31.5×10^{-6} Ω/in.

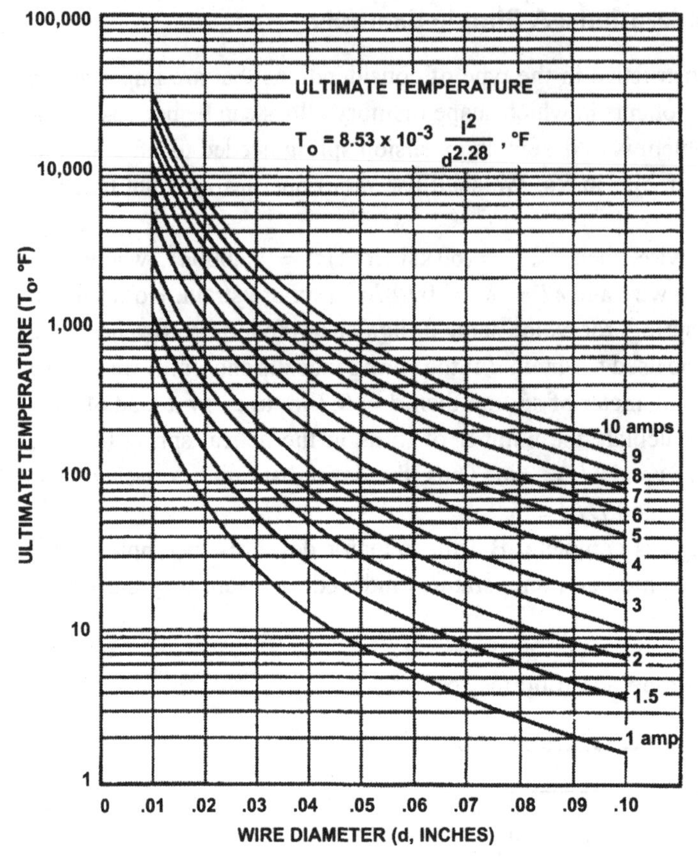

Figure 3.16. SMA spring design chart. (Courtesy of Raychem Corporation.)

allowable limits for the material used. Heating to A_f will lead to full actuation. $A_f = 200\,°\mathrm{F}$ for Tinel. Using $t_H = \ln[1/(1 - (T - T_a)/T_0))]$, one could calculate the maximum temperature to be 891 °F (with T_0 calculated to be 886 °F from the above equation and adding the ambient temperature of 5 °F).

At this stage, some questions of a practical nature can be asked. Will the temperature actually reach 891 °F, considering possible heat losses, and if so should the current be turned off sooner than 5 s? If the current needs to be on all the time, then the temperature should not be allowed to exceed 600 °F. The current can be calculated from the equation $T_0 = (8.532)10^{-3}i^2/d^{2.28}\,°\mathrm{F}$, from which $i = 6\,\mathrm{A}$ is obtained. However, if we were to keep $T_0 = 600\,°\mathrm{F}$, then the required amperage is reduced to 4.8 A in which case the current can be maintained all the time. However, this affects the actuation time. The increased length of time can be calculated from the equation for t_H used earlier and can be shown to be 7.9 s. The question now is if the increased time is acceptable, as the requirement stated earlier was 5 s. Assuming it is acceptable, one can calculate the time to cool from $T_C = T_a + T_s e^{-(t/\tau)}$, where T_s is the temperature from which cooling starts and $T_C - T_a$ is the temperature above ambient to cool to. Because $M_f = 140\,°\mathrm{F}$, cooling will be to a temperature of 135 °F and, therefore, it will take 7.4 s to cool. If

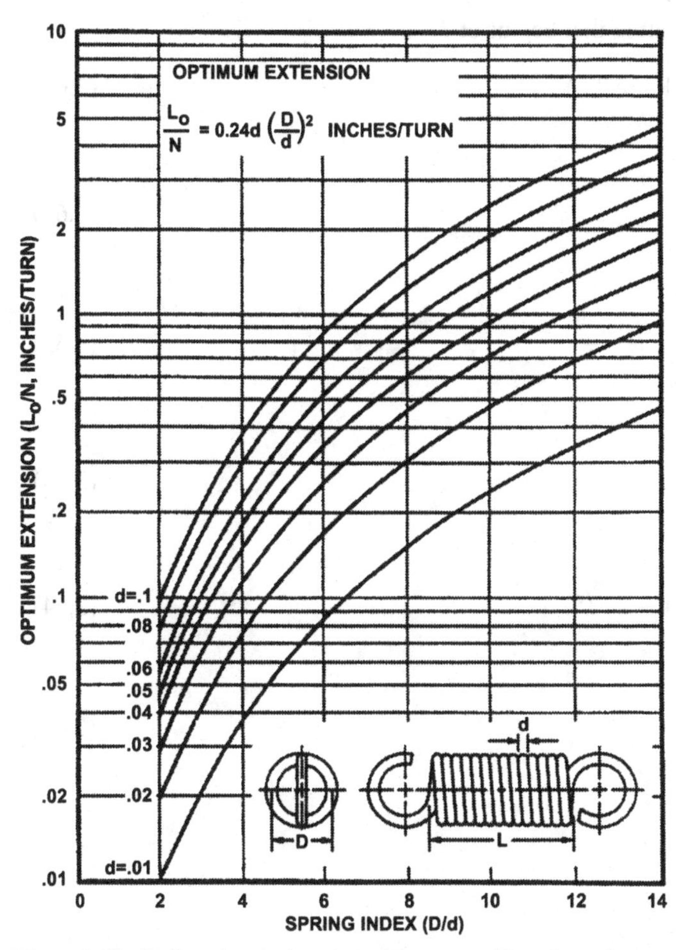

Figure 3.17. SMA spring design chart. (Courtesy of Raychem Corporation.)

this is acceptable, the design is complete; otherwise we need to iterate with the wire size or a different spring index or different material, etc.

Figure 3.18 shows the characteristic temperature and times. The cycle time can be derived as follows:

$$T = T_a + T_0(1 - e^{-t/\tau}). \tag{3.59}$$

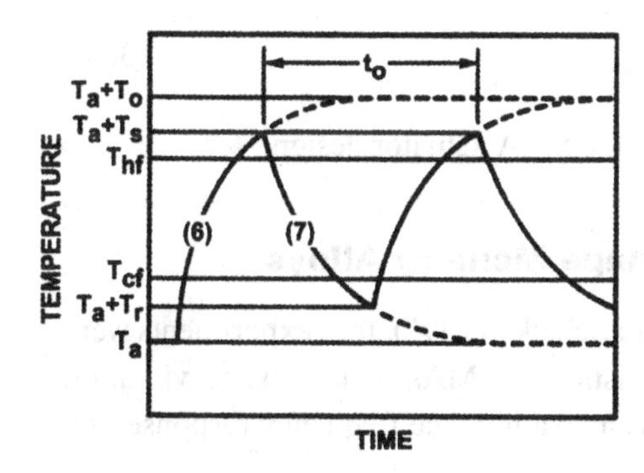

Figure 3.18. Heating and cooling cycles of SMA material. (Courtesy Raychem Corporation.)

The equation is valid for heating that begins at T_a, $a \to b$, $T_a \to T_r$.

$$T_a + T_r = T_a + T_0(1 - e^{-t/\tau}) \tag{3.60}$$

Therefore,

$$\frac{T_r}{T_0} = 1 - e^{-t/\tau} \tag{3.61}$$

and

$$\frac{T_0}{T_0 - T_r} = e^{t/\tau}; \tag{3.62}$$

hence,

$$t_{a\to b} = \tau \ln \frac{T_0}{T_0 - T_r}. \tag{3.63}$$

Heating from $a \to c$,

$$T_a + T_s = T_a + T_0(1 - e^{-t/\tau}) \tag{3.64}$$

and

$$\frac{T_s}{T_0} = 1 - e^{-t/\tau}. \tag{3.65}$$

Therefore,

$$t_{a\to c} = \tau \ln \frac{T_0}{T_0 - T_s}. \tag{3.66}$$

Similarly, cooling from $C \to D$,

$$t_{C\to D} = \tau \ln \frac{T_s}{T_r}. \tag{3.67}$$

The total cycle time is

$$t_0 = t_{ac} - t_{ab} + t_{cd} \tag{3.68}$$

$$= -\tau \ln \frac{T_0}{T_0 - T_r} + \tau \ln \frac{T_0}{T_0 - T_s} + \tau \ln \frac{T_s}{T_r} \tag{3.69}$$

$$= \tau \ln \left(\frac{T_0/T_r - 1}{T_0/T_s - 1} \right). \tag{3.70}$$

This completes all considerations pertinent to the SMA actuator design.

3.5 Vibration Control through Shape Memory Alloys

In the work described by Srinivasan, Cutts, and Schetky (1991), four experiments were carried out to explore some of the characteristics of SMAs of interest to vibration control in structures. The first experiment was to examine the frequency response of a

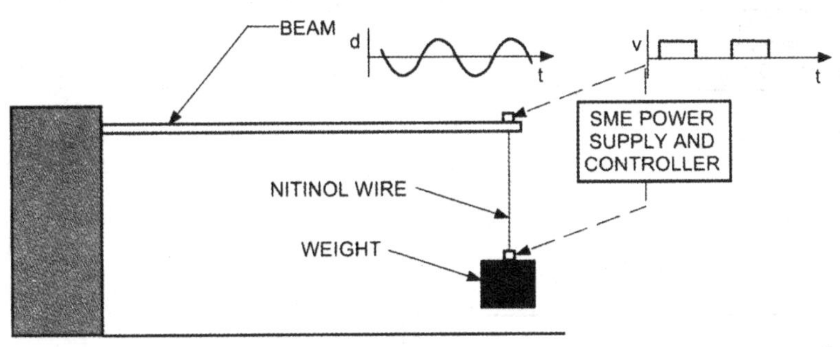

Figure 3.19. Test 1 configuration: cantilevered beam excitation using a NiTiNOL wire-suspended mass system at the free end of the beam.

NiTiNOL wire supporting a weight at the end of a cantilever beam. The experimental apparatus, shown schematically in Fig. 3.19, consists of a steel beam with dimensions $0.42 \times 0.019 \times 0.064$ m from which was suspended a 0.25 kg weight. The suspension wire was 0.152 mm in diameter, which was heated by a square wave generator whose 0.7 A output could be varied over a wide frequency range. The alternating heating and cooling of the wire caused an oscillating force to be imposed on the beam and, at resonance, produced a vibration detectable by touch. Three resonances were observed, with the highest frequency at 168 Hz. The amplitude of the vibrations could be increased by applying forced cooling to the wire.

In the second test, shown schematically in Fig. 3.20, a NiTiNOL wire loop was fastened to a beam of fiberglass-reinforced resin whose dimensions were $0.304 \times 0.0254 \times$

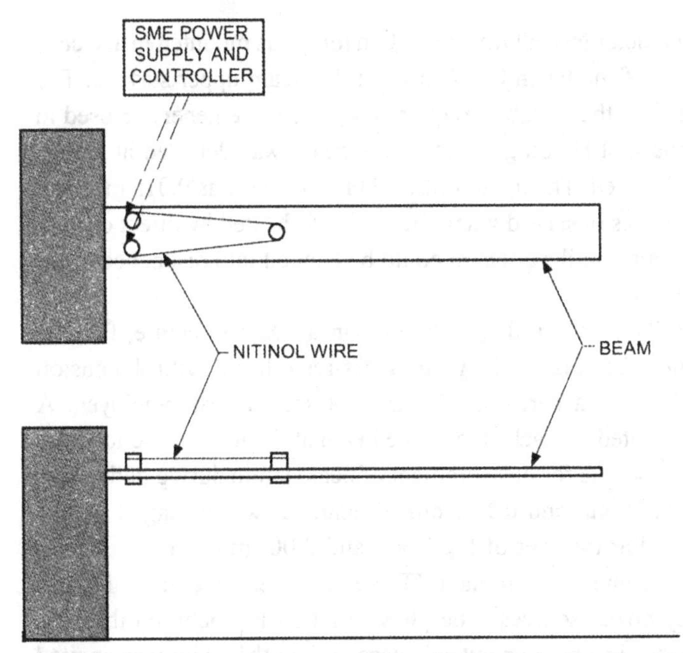

Figure 3.20. Test 2 configuration: cantilevered beam excitation using a NiTiNOL wire forcing element at the root of the beam.

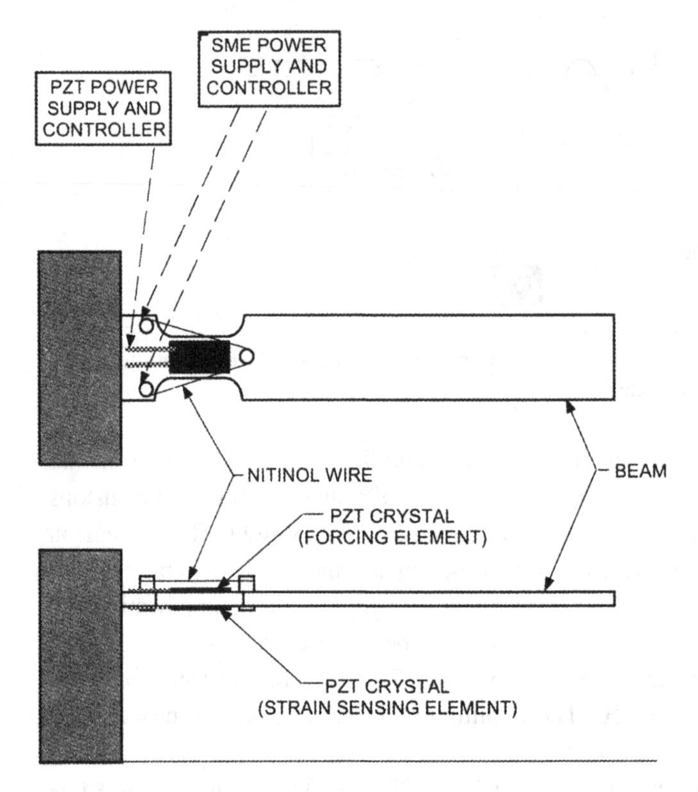

Figure 3.21. Test 3 configuration: vibration control of a cantilevered beam using NiTiNOL wire forcing element at the root of the blade.

0.00159 m. A 0.152-mm-diameter NiTiNOLwire 0.20 m long was mounted in a V configuration on the center axis of the beam 0.006 m from the beam upper surface. The wire was electrically excited by the variable frequency square wave generator used in the first experiment, and the first bending mode of the beam was detected at 10 Hz. No other resonances were detected. The tip amplitude at resonance was 0.015 m, and a static deflection of the beam was observed when the wire was heated by direct current (DC). This demonstrated that a cantilever beam could be excited into resonance by an SMA-generated axial force.

To examine the feasibility of controlling vibration in a beam structure, the test configuration shown in Fig. 3.21 was used. A fiberglass-resin beam with dimension 0.146 × 0.0254 × 0.00159 m and a narrower 0.01 m root section was employed. A piezoelectric crystal was mounted on each side of the beam at the necked section; one for creating a forcing drive, and one for measurement of beam strain during deflection. A NiTiNOLwire 0.076 m in length and 0.152 mm in diameter was arranged in a V configuration axially oriented at the root of the beam and 0.006 mm from the upper surface. Both the SMA wire and the forcing PZT element were energized by two separate variable frequency power sources. The phase relationship between the PZT and wire drive could be either in-phase or out-of-phase. When the beam was excited by the SMA wire heating and cooling, the first mode frequency was observed to be

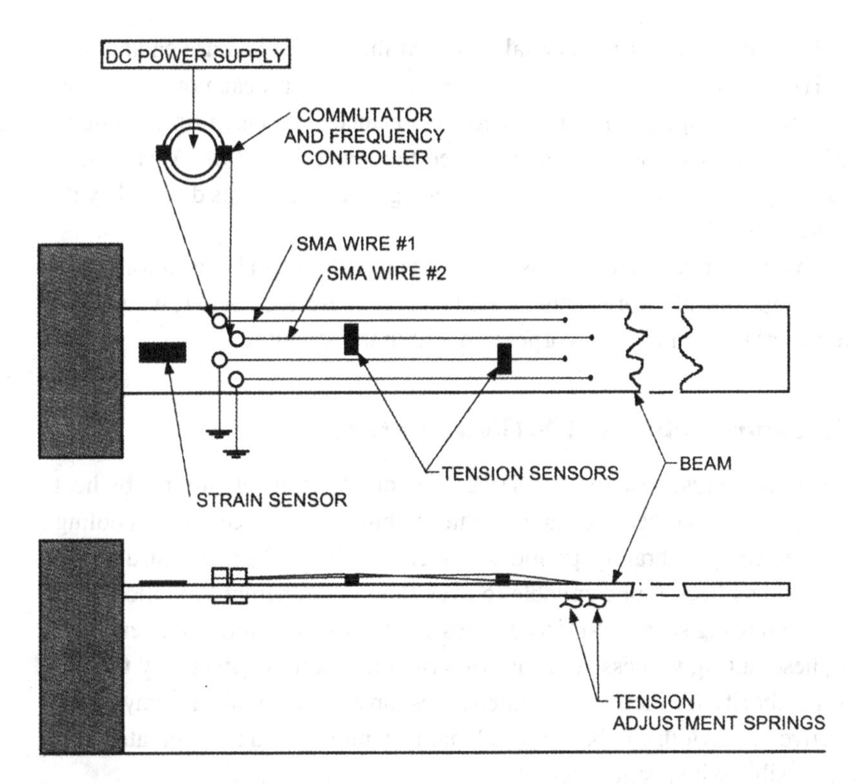

Figure 3.22. Test 4 configuration: cantilevered beam excitation using commutated input signals to multiple NiTiNOL wire forcing elements.

35 Hz, and when excited by the PZT crystal alone, it was 32 Hz. When the PZT crystal was excited at the beam resonance frequency with the SMA wire heated to the point where it was tight (i.e., in tension), the resonance observed was 35 Hz. With the PZT drive set at 32 Hz, the amplitude of vibration was significantly reduced when the SMA wire was tightened by DC heating. This indicated that the system was being detuned by a change in stiffness. When both the PZT crystal and SMA wire were excited in-phase, the amplitude of beam vibration was higher than with either forcing element alone, thus indicating response amplification. The opposite condition prevailed when the PZT crystal and the SMA wire were energized out-of-phase; the amplitude was much smaller, indicating nearly total response attenuation.

In order to demonstrate the concept of frequency bandwidth increase by use of multiple wires energized through a commutator, the experimental arrangement of Fig. 3.22 was employed. Two separate 0.5-mm-diameter NiTiNOLwires were mounted on a fiberglass rectangular box beam whose dimensions were $1.15 \times 0.024 \times 0.0023$ m, with a 0.00125 m wall thickness. The root of the beam had a 0.05 m aluminum insert. The NiTiNOLwires were each 0.56 m long, arranged as two 0.28 m-long pairs, with springs at the free end of the beam to take up wire slack when the wire was cold and to prevent overstraining. The beam resonance was calculated to be approximately 3 Hz, and the wires were energized through a commutator at half this frequency. The dynamic tension of the wires was measured with a sensor placed on each pair, and the beam strain was measured by a PZT strain measuring crystal placed at the root. When driven at a half frequency (1.5 Hz), the wire tension signals are out-of-phase with each other, and an effective 3 Hz drive was apparent from the beam resonance. The beam tip amplitude was about 0.013 m. This successfully demonstrated the feasibility of SMA excitation of a structure at frequencies above the maximum for a given wire size as dictated by its thermal response.

These experiments demonstrate various ways in which SMA can be used for active and passive vibration control. More sophisticated tests are described below, which will use composite materials with SMA incorporated within the structure.

3.6 Multiplexing Embedded NiTiNOL Actuators

In many applications the response time of SMA actuators is ultimately limited by heat transfer. Typically, very rapid heating can be achieved but the time needed for cooling is "long" compared to the vibratory period of typical mechanical or civil structures. A unique approach has been found suitable to overcome this inherent limitation. The approach is based on using several NiTiNOLwires as actuators in parallel and energizing subsets of these during successive cycles of structural motion, effectively trading reduced control authority for increased frequency response. Thus, with an array of actuators an effective bandwidth can be achieved that is demonstrated to be greater than the bandwidth possible with a single actuator.

This technique of multiplexing several actuators was established in principle as described above. The approach was extended to demonstrate its validity by applying it to a robust box beam made of steel. Furthermore, successful performance of the multiplexing scheme has led to the analysis, design, fabrication and testing of a composite beam in which NiTiNOLfibers were embedded (Srinivasan et al., 1997). A series of vibration tests were conducted on the composite beam along with temperature measurements using an infrared camera. With the multiplexing approach, the first two modes of the composite beam, at 23.5 Hz and 144 Hz respectively, were excited.

With the use of an array of actuators in parallel, the relationship of the signal driving any one actuator, taken to be a rectangular pulse of period T_e, to the desired structural response, of period T_r, is shown in Fig. 3.23. For purposes of discussion of the concept, four actuators in parallel have been assumed here, but the method is applicable to any number. The excitation has a short duty cycle (ratio of time "on" to pulse length), with the "on" period equal to approximately $T_r/2$, i.e., the portion of a cycle of the response during which the structure is moving in one direction (shown as the positive-going half in the figure). The corresponding actuator responds quickly to heating, producing an increasing force during the "on" pulse, then cools and relaxes during the succeeding $3\frac{1}{2}$ cycles of structural response. No actuator is driven during the negative-going half of the first structural response cycle of Fig. 3.23, but at the end of time T_r, an identical pulse is applied to the second of the four actuators. This process is repeated with the remaining two actuators, so that after all four have been switched on for $T_r/2$ and allowed to

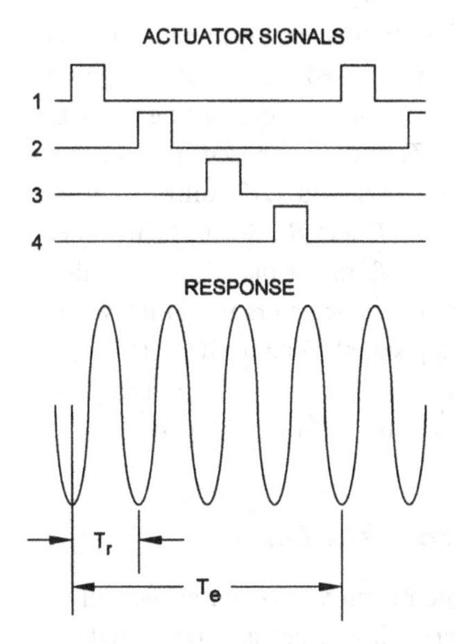

Figure 3.23. Theoretical timing of SMA excitation and beam response.

cool for $7T_r/2$, the first is again activated. The force experienced by the structure is therefore the superposition of the four actuator outputs, with period $T_r = T_e/4$; that is, the time-varying force driving the structure has a frequency four times that of the electrical signal driving each actuator. More generally, the force experienced by the structure will be of a frequency equal to the electrical driving frequency multiplied by the number of actuators in parallel.

Both of the test specimens discussed here incorporated four actuators, each consisting of one or more SMA fibers electrically connected in series. Passing current through the wires of an actuator produced a bending moment that deflected the cantilevered beam specimen toward the actuator, that is, up or down, depending on the whether the actuator wire was nearer to the top or bottom face of the beam.

A nomenclature of "firing" (driving) an actuator and a "firing order" for the actuator array was adopted, along with a convention for describing the signals used. Those actuators arranged so as to pull the beam up are denoted by T_i, and those pulling it down by B_i. The steel beam specimen described below had four independently controlled wires mounted above the beam and none below, and hence was driven by four actuators in parallel, denoted T_1 through T_4. The composite beam had two independently controlled sets of wires in its top face, comprising actuators T_1 and T_2, and two sets in its bottom face, actuators B_1 and B_2. The firing order for the actuator array is described for one period T_e of the excitation by stipulating the actuators to be fired during each half-cycle of a structural response period T_r separated by commas, with semicolons used to separate the groups for consecutive periods T_r. Zero is used as a place holder to indicate that no actuator is energized for one half-cycle of structural motion. For example, the excitation shown in Fig. 3.23 would be denoted T_1, 0 and the superposition of four such signals applied to the wires atop the box beam would be written T_1, 0; T_2, 0; T_3, 0; T_4, 0. Some additional complexity is introduced when actuators oppose one another and can act during successive half-cycles of structural motion. In this arrangement, actuators are embedded in both faces of the beam to produce a coordinated action with the top set pulling the beam up during the first half cycle and the bottom set next pulling it down. Considering the composite specimen with actuators T_1 and T_2 opposed by actuators B_1 and B_2, we can have, for example, T_1 pulling up for a period $T_r/2$ followed by B_1 pulling down for $T_r/2$, with this pattern then repeated by T_2 and B_2. Such a firing order is denoted $T_1, B_1; T_2, B_2$. Note that this implies $T_e = 2T_r$ rather than $T_e = 4T_r$, that is, the frequency of structural excitation available from the actuators used this way is twice rather than four times that of the electrical driving signal. For a fixed total number of actuators, less frequency gain can be achieved if the structure is driven during the entirety of each response cycle rather than during only one-half of each cycle.

3.6.1 Analytical Basis for the Design of Composite Beams

In order to design a composite specimen with shape memory alloy wires as well as structural fibers embedded in an epoxy matrix material, it is necessary to account for the effects of the SMA on the mechanical properties of the composite and for its role as an actuator. Because the volume fraction of SMA, at least within some laminae, may be large, it is natural to treat the wires as inclusions in a matrix of composite material whose properties are known. This allows the use of existing formulas to estimate the influence of SMA fraction and placement on bending and in-plane stiffnesses and thus on the specimen's natural frequencies of free vibration.

In the case of a one-dimensional structure, such as the beam specimen of interest here, significant simplifications are possible. By considering simultaneously the geometry and material properties of the laminae, Gibson and Plunkett (1976) derived an equivalent modulus for a composite beam in bending. This modulus can be used directly in formulas previously derived for homogeneous, isotropic beams. This equivalent modulus aids physical insight as well as computational convenience. For a beam symmetric about its midplane and with an odd number n of laminae, the equivalent modulus is given by

$$E_c = \frac{8}{n^3}\left[\frac{E_0}{8} + \sum_{j=1}^{(n-1)/2} E_j(3j^2 - 3j + 1)\right], \tag{3.71}$$

where E_j is the modulus, in the longitudinal direction, of the jth lamina. These properties were readily available for the prepreg materials under consideration at the time the composite specimen was being designed.

Major features of the beam cross section are shown schematically in Fig. 3.24. The presence of NiTiNOL wires in some of the beam laminae was taken into account by computing an effective modulus for those plies using a rule of mixtures based on the volume fractions of SMA and prepreg material. Because the equivalent modulus equation above assumes all the beam's laminae are of the same thickness, the foam

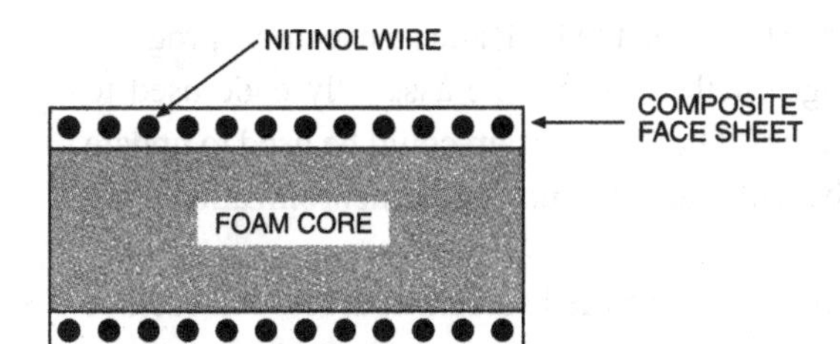

Figure 3.24. Cross section of composite beam with embedded NiTiNOL wires.

core was modeled as a number of plies each as thick as one prepreg strip but with the isotropic material properties of the much more compliant foam.

3.6.2 Control Scheme and Test Specimens: Analysis and Design

Experiments were carried out on two specimens: a steel box beam with externally mounted SMA wires, and a composite beam with NiTiNOL wire embedded in its outer face sheets. This section discusses the construction of the steel beam, the design of the composite specimen fabricated by Martin Marietta, and the mounting fixture used to secure each specimen as a cantilevered beam. The control software and hardware used to generate and amplify the electrical driving signals were common to all the experiments, and are described first.

3.6.2.1 Electrical Control System

A schematic of the test setup is shown in Fig. 3.25. Voltage signals were generated by using a digital signal processing (DSP) card installed in a Pentium PC. The DSP card, from Spectrum Signal Processing, is based on a Texas Instruments TMS320C30 integrated circuit and can be programmed in a high-level language. Code written on the PC is cross-compiled, then downloaded to the DSP card and executed there under the control of a user program running on the PC. Input and output (I/O) to and from the DSP is in the form of analog signals in the range ±2.5 V. Two channels of I/O were available directly on the DSP card and two more were added through an expansion card, giving a total of four independent signals available to drive the test apparatus.

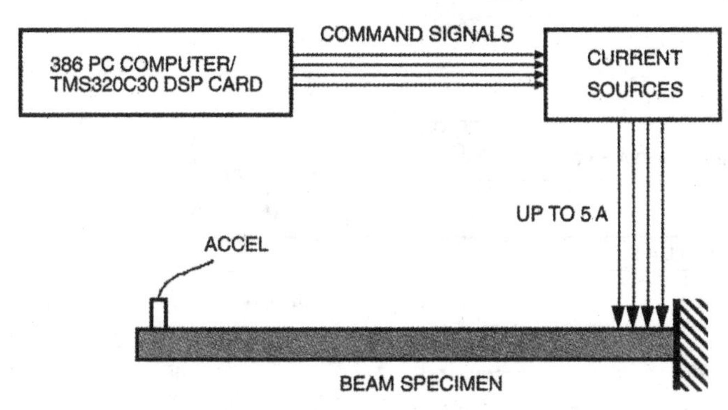

Figure 3.25. Block diagram of electrical system.

A signal generator program was written to run on the DSP and an executive program to run on the PC, both in the C language (with some in-line assembly code used to control low-level DSP operations and status). The PC program could be used to update the output levels and frequency on the fly, with pulse shaping and synchronization done automatically.

The DSP output signals drove four voltage-controlled current sources, and each of these in turn powered one set of SMA wires in series (one actuator). The current sources had a sensitivity of 2.0 A/V and could apply up to 5 A or 20 V DC to a resistive load before overheating or saturating.

Although the DSP outputs are extremely stable, the finite rate of analog-to-digital conversion has the effect of discretizing the available output frequency spectrum. Most tests reported here were performed with a conversion frequency in the range of 1 kHz to 10 kHz. Although the resulting precision was much better than could be attained with an analog signal generator, the steps in driving frequency are reflected in some of the response plots given later.

3.6.2.2 Preliminary Tests: Steel Beam with External Actuation

To produce a specimen for preliminary experiments at low cost, SMA wires were attached to a steel box beam. Four strands of NiTiNOL wire were mounted on the top of the horizontally cantilevered beam. The wires were supported by passing them through Teflon blocks clamped to the beam. Stainless steel collars were crimped to the wires on the outside of the Teflon blocks, that is, toward the beam ends. The blocks were moved apart until the wires were taut at room temperature, and the tension of individual wires was adjusted by inserting split washers between the stainless steel collars and the outer face of the restraining blocks. Figure 3.26 shows the geometry and dimensions of this specimen.

The steel beam is shown installed in its massive, rigid base in Fig. 3.27. In that photograph, taken during an early stage of testing, several temporary loops of SMA and current supply wires can be seen protruding from the beam. Prior to the experiments

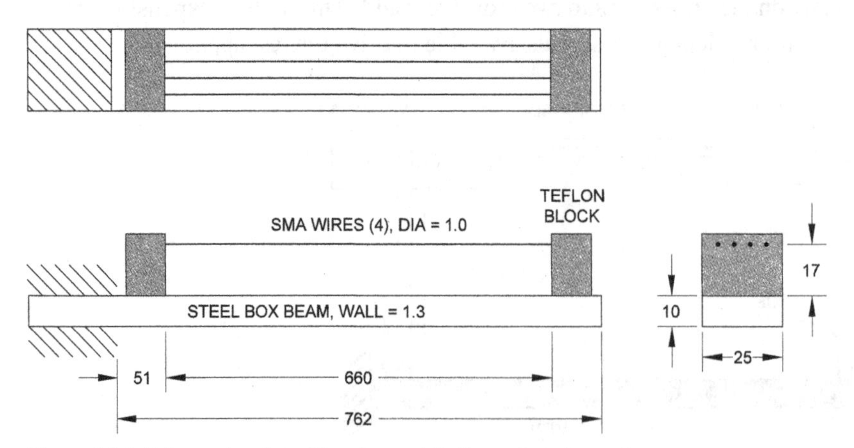

Figure 3.26. Schematic of steel box beam (all dimensions in mm).

Figure 3.27. Box beam in test fixture.

reported in the following sections, these connections were made permanent, the copper lead wires were anchored to the structure, and the leads from the beam-tip ends of the SMA wires were brought back through the box beam to its root.

Each of the four NiTiNOL wires mounted on the beam had a nominal diameter of 1.0 mm and exhibited a resistance of 1.8 Ω, compared to which the resistance of the copper leads was negligible. Each wire was driven by one channel of the electrical control system described above. The wire was characterized by differential scanning calorimetry, and the following transformation start and finish temperatures were determined: $A_s = 38.6\,°\text{C}$, $A_f = 55.0\,°\text{C}$, $M_s = 27.8\,°\text{C}$, and $M_f = -1.6\,°\text{C}$. It is clear that the austenitic transformation may be accomplished by resistive heating, but that only a partial reversion to martensite can be expected because normal laboratory room temperature is just a few degrees Celsius below the measured M_s.

3.6.2.3 Composite Beam with Embedded Fiber Actuation

On the basis of the equivalent modulus equation for bending, several sets of design curves were computed by using a simple Matlab program written for the purpose. It

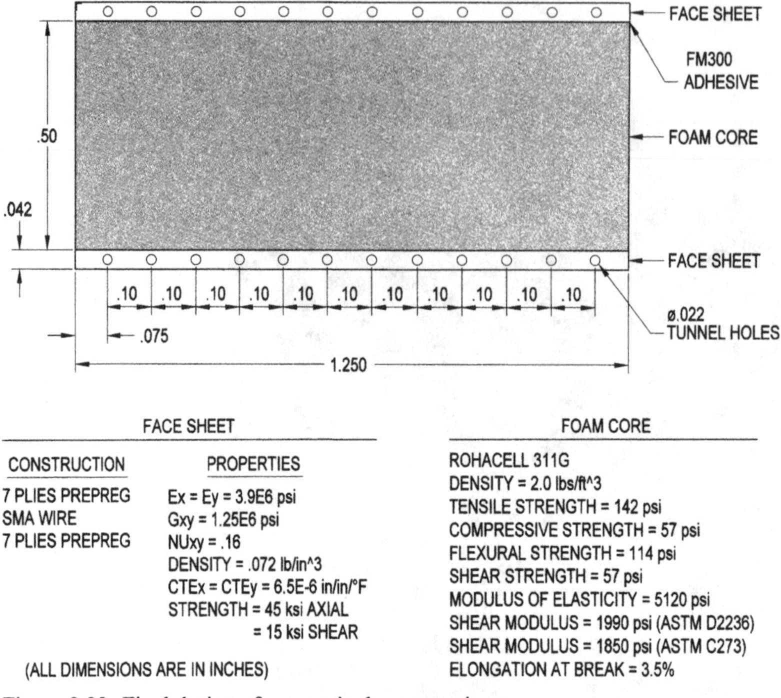

Figure 3.28. Final design of composite beam specimen.

was desired to design a specimen whose first natural frequency was high enough, greater than approximately 10 Hz, to allow the use of existing dynamic test equipment, but not so high that significant heating and cooling of the embedded SMA could not take place during one cycle of vibration. Beam dimensions and layup were selected, and the final design shown in Fig. 3.28 was developed in conjunction with engineers and technicians at Martin Marietta. Calculations made on the basis of the equivalent modulus previously presented predicted the first natural frequency of the specimen would be approximately 35 Hz.

The beam consisted of a rectangular foam core to which were bonded two composite face sheets consisting of fourteen plies of glass-epoxy composite. The sprung length of the beam was 559 mm; other dimensions and material properties are given in Fig. 3.28. Embedded in each face sheet were twelve strands of NiTiNOL wire, evenly spaced and parallel to the beam axis. The wire was of 0.020 in. (0.5 mm) diameter and was installed with a 3% tensile prestrain. These were connected in a total of four electrical circuits, one to be driven by each channel of the control system. Thus, with the beam mounted as shown in Fig. 3.29, there were two sets of six wires each in the top face sheet and two sets in the bottom, each set of six wires acting as a single actuator. The wires of each set were selected to be symmetric about the vertical plane of symmetry of the specimen, so that activating any one set of wires would produce bending but not torsion of the beam. Material properties of the embedded wires are summarized in Table 3.5.

Figure 3.29. Composite beam with foam core in test fixture.

When the beam was received, the insulated leads to the four sets of wires had been stripped and tinned. Lugs were soldered to these leads, which came out at the beam root, and they were connected to a nearby terminal block. At each end of the beam there were connections between the SMA wires of each set, made with small loops of bare brass wire and insulated with Gliptol. Resistance measurements showed this insulation to be inadequate, and an additional layer of high-temperature RTV silicone was added, encasing each run of adjacent loops. Before this was applied, the loops of wire to be insulated were separated and splinted; after the silicone had cured, the splints were removed. This eliminated any electrical interaction of the actuator circuits. The six wires in series comprising one actuator had a resistance of approximately 18 Ω.

3.6.2.4 Steel Beam with External Actuation

Preliminary testing of the steel beam with an instrumented hammer, while the attached SMA wires were at room temperature, showed that several modes existed at frequencies

Table 3.5. Material Properties of NiTiNOL Wire Embedded in the Composite Beam Specimen

Diameter	0.51 mm (0.020 in.)
A_f	57 °C
E (martensite)	1.1 MSI
E (austenite)	7 MSI
Prestrain (as installed)	3 %

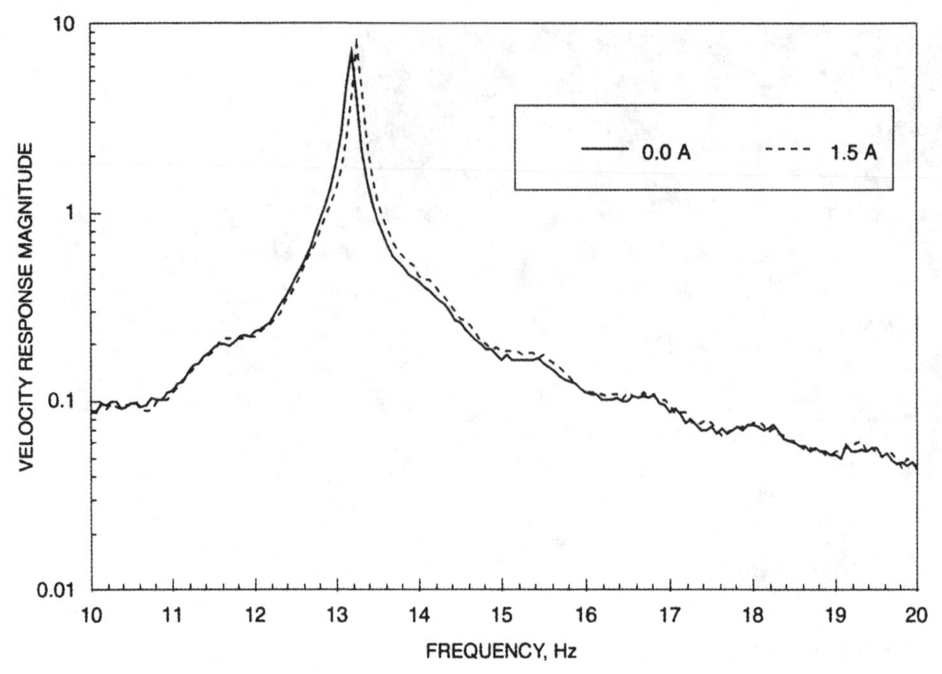

Figure 3.30. Comparison of FRFs of steel beam at constant currents of 0.0 A (room temperature) and 1.5 A.

below 1 kHz. Because the bandwidth of the SMA excitation was expected to be limited, attention was focused on only the first two modes.

The FRF data over the range of 10–20 Hz, which encompasses the first natural frequency of the structure at two current levels (0.0 and 1.5 A), were extracted and are plotted in Fig. 3.30 to better show the influence of steady heating on the response. The effect is quite small, despite continuous heating of the wires producing a clearly visible deflection of the tip of the cantilevered beam of the order of 3 mm, compared to its room temperature position. That this change in the dynamic response is slight is due to the high axial stiffness of the steel box beam (put in compression by the heating-induced contraction of the wires), and to the SMA wires on this specimen being anchored approximately 5 cm from the root of the beam. The greatest bending stresses produced by the hammer blow were at the root of the beam, and because the wires do not span this section of the beam they do not experience strains as great as if they were anchored at the root.

With the fundamental frequency of the steel beam specimen known from modal testing, the electrical controlling hardware was programmed to produce four signals in the form of rectangular pulses with period four times that of the desired beam motion and staggered by one-fourth of this period. The pulses had a duration equal to one-eighth of their period. Consequently, current was applied to one of the four SMA actuator wires during the positive-going half of each cycle of structural motion, but a given wire was energized only during one-half of every fourth cycle of this motion. In the nomenclature previously introduced, the firing order was T_1, 0; T_2, 0; T_3, 0; T_4, 0 with a pulse duration (duty cycle) of $T_e/8$. The amplitude of the control signals was adjusted so that the current regulators would produce a current of 5 A in their respective wires during the "on" portion of the driving signal and no current during the "off" portion. Room temperature air was blown over the wires, along the beam, during this test.

The period of the excitation was then varied so that the frequency of the forcing experienced by the structure was swept from 13.10 Hz to 13.30 Hz in steps of 0.02 Hz. The beam tip accelerations thus induced were recorded as previously described. These were plotted and a peak response frequency of 13.24 Hz and half-power bandwidth of 0.06 Hz were estimated, and a damping ratio of 0.2% was then computed.

With the excitation frequency and waveform fixed, the electrical energy input to the actuator wires, and thus the heat dissipated, varies as the square of the driving signal level (assuming, as was found to be true of this specimen, that electrical resistance remains nearly constant during temperature and phase changes). Obviously, some current is necessary to produce a response, but an unnecessarily large excitation current could result in the SMA actuator temperature remaining above M_s during the entire "off" portion of the excitation, greatly reducing the effectiveness of the actuator. Therefore there will be an optimal amplitude for the driving current, and an experiment was designed to find this value.

The driving frequency was fixed at 13.24 Hz, the frequency of maximum response found previously. The excitation pulse amplitude was adjusted to 2, 3, 4, and 5 A and beam tip acceleration recorded. The results are plotted in Fig. 3.31. From inspection of this curve, it was concluded that the 5 A drive level, the maximum the control hardware

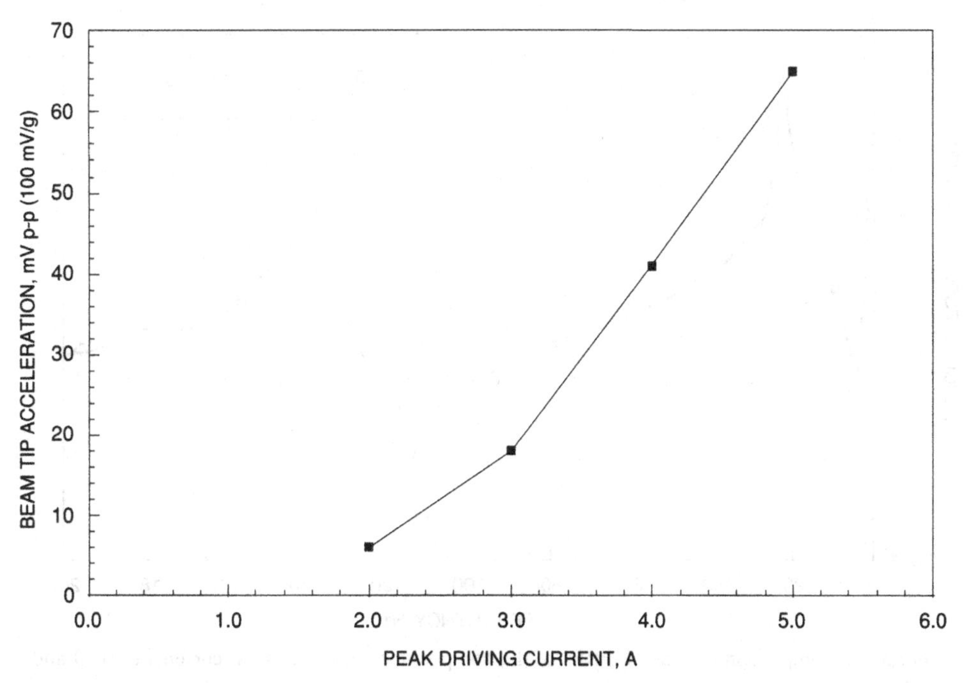

Figure 3.31. Steel beam Mode 1 tip acceleration as a function of peak current in SMA wires.

was capable of, was not producing temperature saturation as described above, and this level was used for all subsequent tests on this specimen. In fact, it appears that larger driving currents could be used effectively, were they available.

3.6.2.5 Composite Beam with Embedded NiTiNOL Fibers

For each of the three stages of testing, the composite beam was mounted in the test fixture as previously described. Hammer tests were carried out at each stage to identify the first and second natural frequencies and damping ratios at room temperature and with various constant currents flowing in the NiTiNOL wires. The first natural frequency of the specimen was measured as 23.5 Hz, compared to approximately 35 Hz predicted during design. This difference of 33% is attributed largely to mass of the wiring and insulation at the free end of the beam, which was not modeled in the design. The 23.5 Hz fundamental frequency was well above the 10 Hz minimum considered while planning the experiments. Pulsed electrical excitation of the SMA was then employed to drive the structural response, and beam tip acceleration was recorded as for the steel beam specimen.

In Stage 2, hammer tests were conducted on the composite specimen while constant currents of 0.25 and 0.50 A were supplied to its NiTiNOL actuator wires. The FRFs obtained at room temperature and at a current of 0.50 A are plotted simultaneously in Fig. 3.32. The response in the first mode was virtually unaffected by this heating, but the second mode response amplitude was significantly increased. Natural frequency and

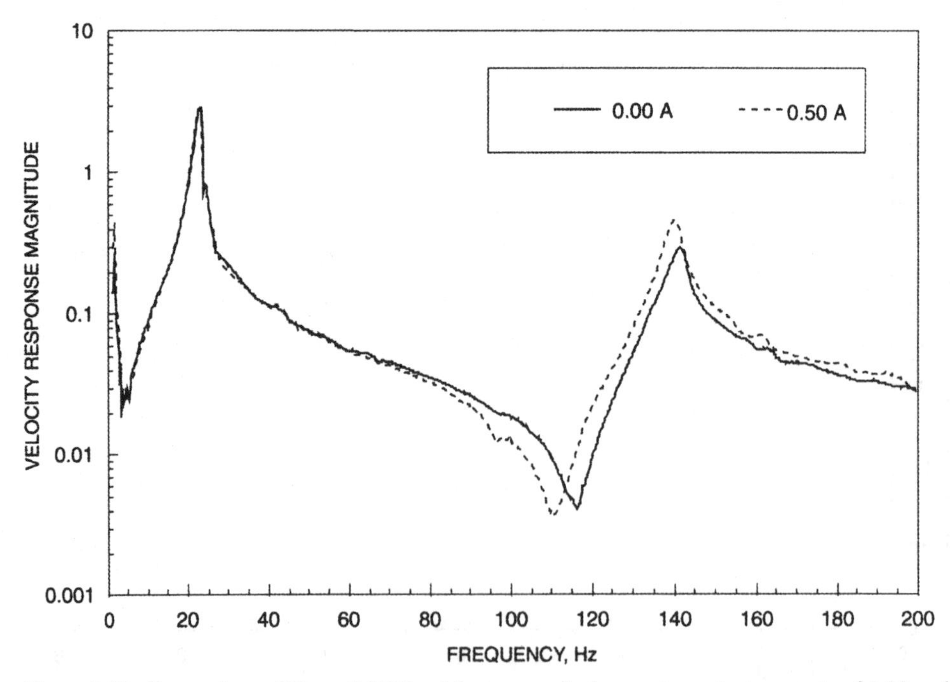

Figure 3.32. Comparison of Stage 2 FRFs of the composite beam at constant currents of 0.00 and 0.50 A.

Table 3.6. Natural Frequency and Damping Ratio Estimates from Hammer Tests of the Composite Beam with Embedded NiTiNOL Wires

		Constant Current, A			
		0.00	0.25	0.50	0.75
Stage 1:	f_1, Hz	22.5	—	—	—
	ζ_1, %	4.2	—	—	—
	f_2	139.30	—	—	—
	ζ_2	1.5	—	—	—
Stage 2:	f_1	22.20	22.01	21.96	—
	ζ_1	5.1	4.9	5.1	—
	f_2	140.30	139.11	139.29	—
	ζ_2	1.9	1.6	1.6	—
Stage 3:	f_1	23.47	23.51	23.50	23.49
	ζ_1	1.9	1.9	1.6	1.9
	f_2	144.01	143.28	142.34	141.58
	ζ_2	1.2	1.4	1.4	1.6

damping ratio estimates obtained from the constant-current FRFs are given in Table 3.6. Similar results were found from the Stage 3 hammer tests, and are included in Table 3.6.

This time an additional constant current, 0.75 A, was used. The FRFs from this stage of testing are compared in Fig. 3.33 for currents of 0.00 and 0.50 A. Note that the

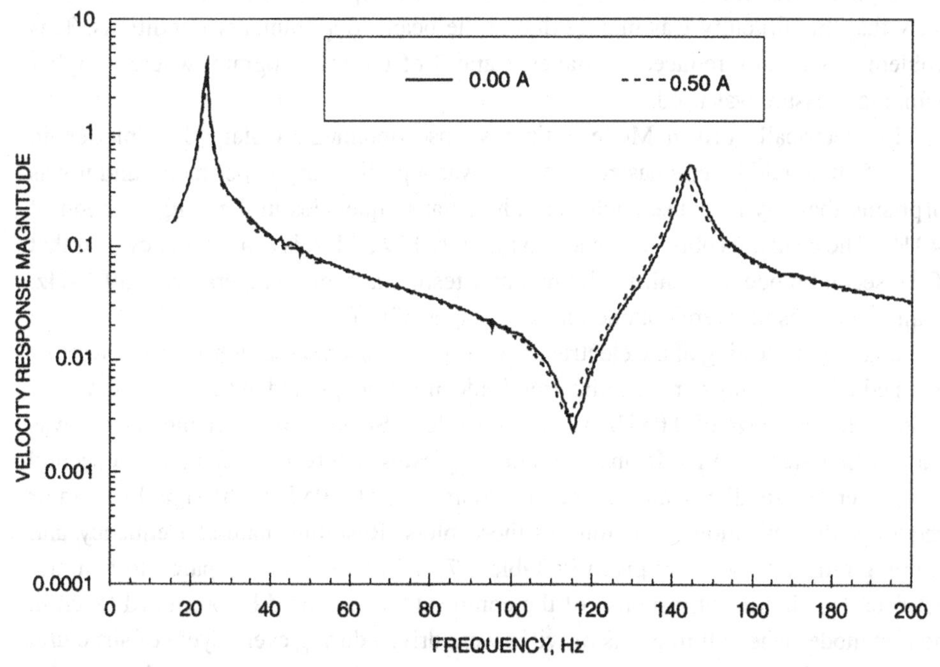

Figure 3.33. Comparison of Stage 3 FRFs of the composite beam at constant currents of 0.00 and 0.50 A.

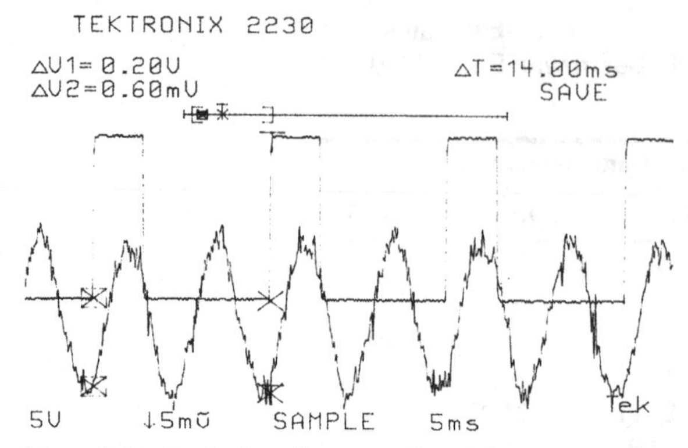

Figure 3.34. Excitation of actuator T_1 and tip response of composite specimen.

heat dissipation resulting from the 0.50 A current flowing in all four sets of actuator wires was the same as was produced by the 1.0 A current driving only one wire at a time that was used in the tests of the electrically driven response, discussed below. The effect of constant heating is less pronounced here than it was in Stage 2, but the FRFs of the structure at 0.50 A are quite similar in the two stages.

To drive the composite beam electrically at its fundamental frequency, an actuator firing order of T_1, B_1; T_2, B_2 was used, and preliminary experiments led to the selection of 1 A as the driving current pulse amplitude. Corresponding oscilloscope traces are shown in Fig. 3.34. The FRF was repeatable but not easily interpreted, and so testing was stopped without identifying a global maximum response frequency. It now seems likely that the difficulty was mainly due to the beam root boundary conditions. This problem was greatly reduced in Stages 2 and 3 of the test program, where a higher clamping pressure was used.

The electrically driven Mode 2 tip response obtained in Stage 1 is plotted in Fig. 3.35. It should be emphasized that this was a preliminary experiment, and it was surprising that any response could be induced at frequencies in the neighborhood of 140 Hz. The peak thus obtained, at approximately 139.35 Hz, is at a frequency near that of the second mode as identified by hammer testing at room temperature, 139.30 Hz. Again, better results were obtained in later stages of testing.

Stage 2 and 3 testing of the electrically driven response was straightforward. Results obtained with a driving current pulse amplitude of 1 A are plotted in Figs. 3.36 and 3.37. In Stage 2, a step size of 0.05 Hz was used, while in Stage 3 one frequency sweep was made with a step of 0.10 Hz and the area of greatest interest selected, and a second sweep over this smaller band was then made in steps of 0.05 Hz. The signal generator frequency discretization is obvious in these plots. Resulting natural frequency and damping ratio estimates are given in Table 3.7, where they are compared to hammer test data. No significant response of the composite beam could be produced (even in the first mode) when all four sets of wires were driven during every cycle of structural motion; that is, multiplexing allowed effective control to be achieved where it was otherwise impossible.

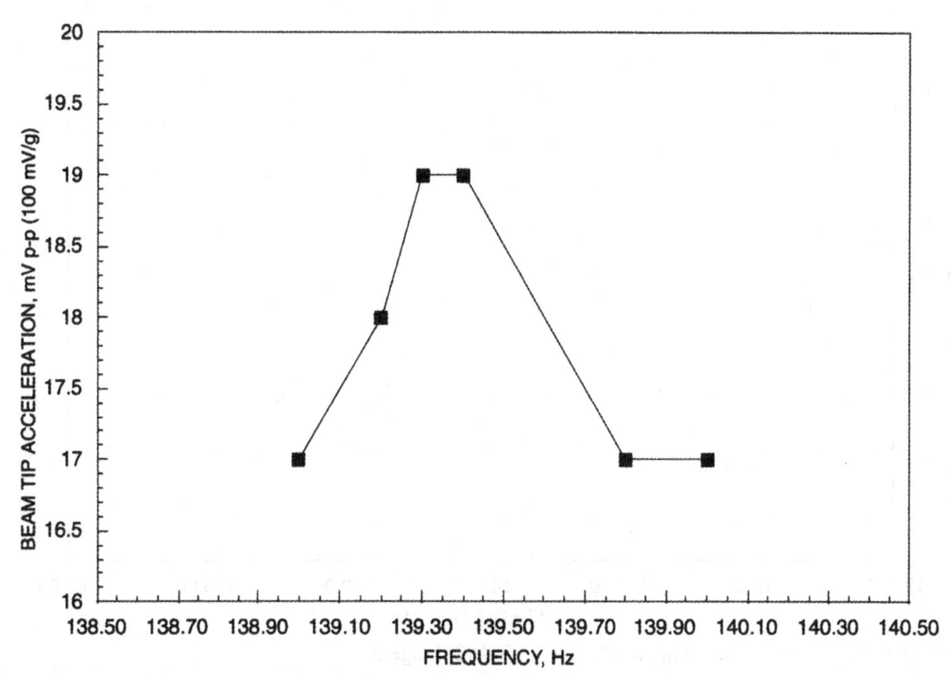

Figure 3.35. Composite beam tip acceleration, Mode 2, Stage 1.

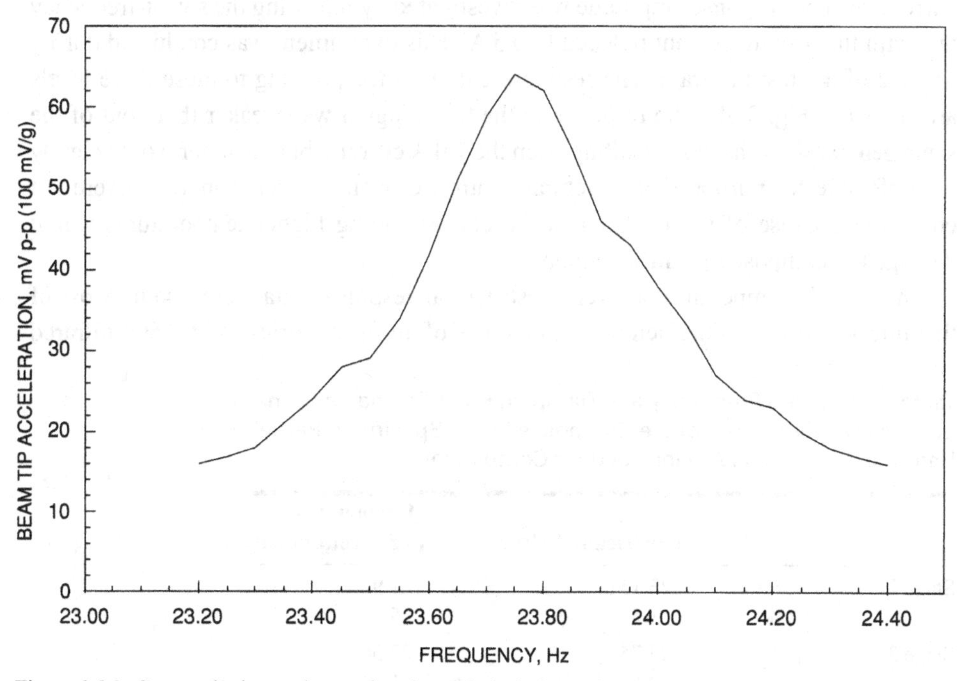

Figure 3.36. Composite beam tip acceleration, Mode 1, Stage 3.

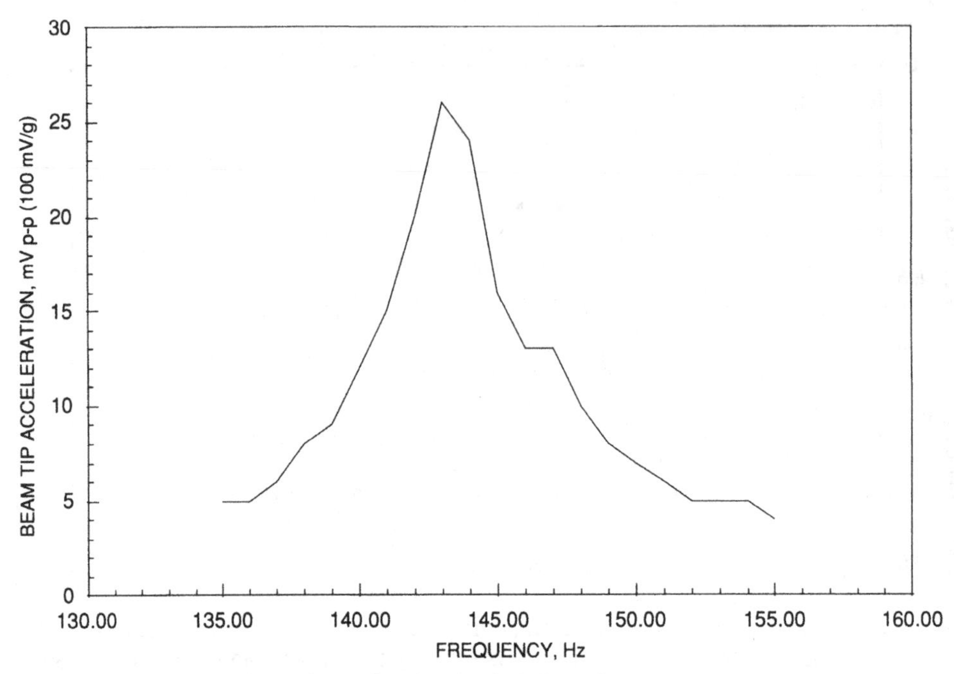

Figure 3.37. Composite beam tip acceleration, Mode 2, Stage 3.

Once it was determined that a significant tip response could be produced near the first natural frequency by an electrical driving current of 1.0 A amplitude, the effect of the current level on response amplitude was investigated by repeating the swept-frequency test with the driving current reduced to 0.5 A. This experiment was conducted during Stage 2 of the test program. The response curves corresponding to these drive levels are shown in Fig. 3.38. The response to the 0.5 A signal was weaker than, but of the same general shape as, that resulting from the 1.0 A current, but the latter would clearly be preferable for purposes of structural control. Currents larger than 1.0 A were not employed because of the danger that the corresponding higher temperatures would damage the composite beam specimen.

In Stage 3, temperature as well as structural response data were taken. Most of this information was obtained through the use of an Inframetrics Model 600 infrared

Table 3.7. Natural Frequency and Damping Ratio Estimates from Electrically Driven Tests of the Composite Beam Specimen (Partial Hammer Test Results Are Included for Comparison)

		Electrically Driven	Hammer Test (0.5 A continuous)
Stage 2	f_1, Hz	23.15	21.96
	ζ_1, %	1.5	5.1
Stage 3	f_1	23.75	23.50
	ζ_1	0.6	1.6
	f_2	144.70	142.34
	ζ_2	0.9	1.4

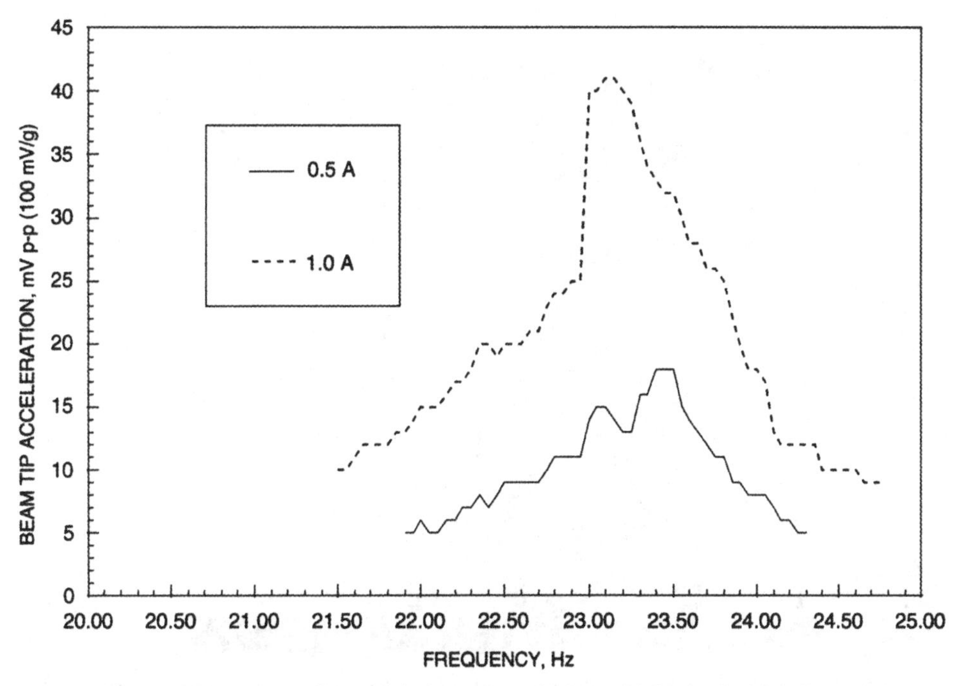

Figure 3.38. Stage 2 response of composite beam at peak currents of 0.5 A and 1.0 A.

camera, which provided real-time video images of the temperature field of the specimen. Because the camera uses liquid nitrogen for a temperature reference, there are limits on its orientation. Therefore, all thermal tests were done with the beam and mounting fixture rotated so that the beam axis was vertical. Temperature at a point could be read by positioning a cross-hair cursor controlled through the camera's software. The system was calibrated by mounting a type K thermocouple to the beam and using its output as a reference to correct the raw numbers available from the camera. The thermocouple is visible against the background thermal image of the beam in Fig. 3.39, which was captured from video tape using NIH Image software.

The beam temperature profile at a fixed longitudinal position was measured by stepping the cursor across the face of the beam to ten evenly spaced stations. During these measurements, the beam was driven electrically at Mode 1 resonance with a pulse amplitude of 1.0 A. The temperature values recorded are plotted in Fig. 3.40. Edge effects were evident, as was expected, and the symmetry of the profile about the beam centerline was also apparent.

Assuming uniform, one-dimensional heat flux and considering the symmetry of the specimen, it was found that a current of 1.0 A flowing in one of the four NiTiNOL actuators at any time produced a mean flux through the composite face sheets of 0.05 W/cm^2. The same flux was achieved during hammer testing by supplying 0.50 A to all four sets of actuator wires simultaneously. It may be noted that this is roughly one-fifth the continuous heat flux typically used, for example, during the curing of field-applied composite patches.

It was observed that the matrix material of the face sheets felt rubbery at the temperatures reached in steady-state operation, indicating that the glass transition temperature

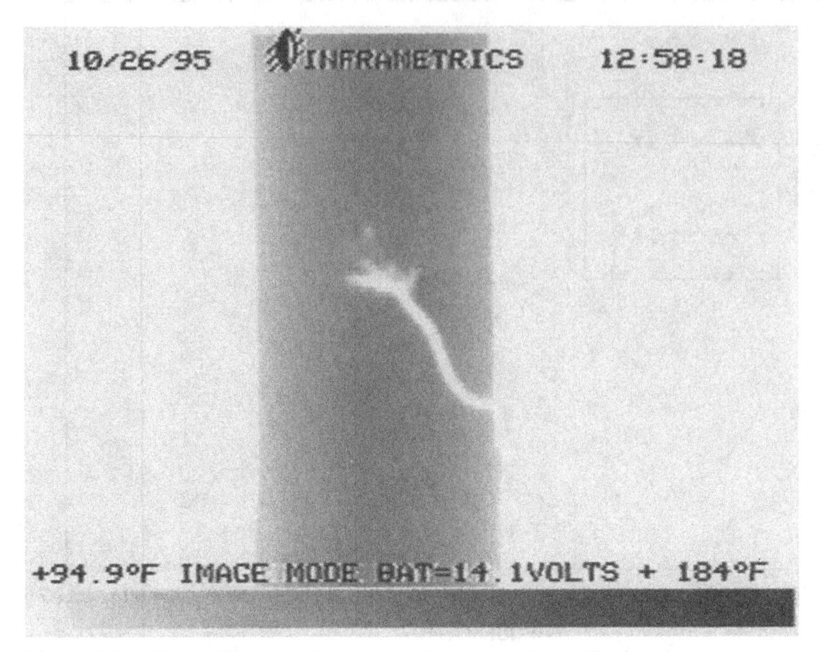

Figure 3.39. Thermal image of composite beam specimen (thermocouple temperature reference is visible against the beam face).

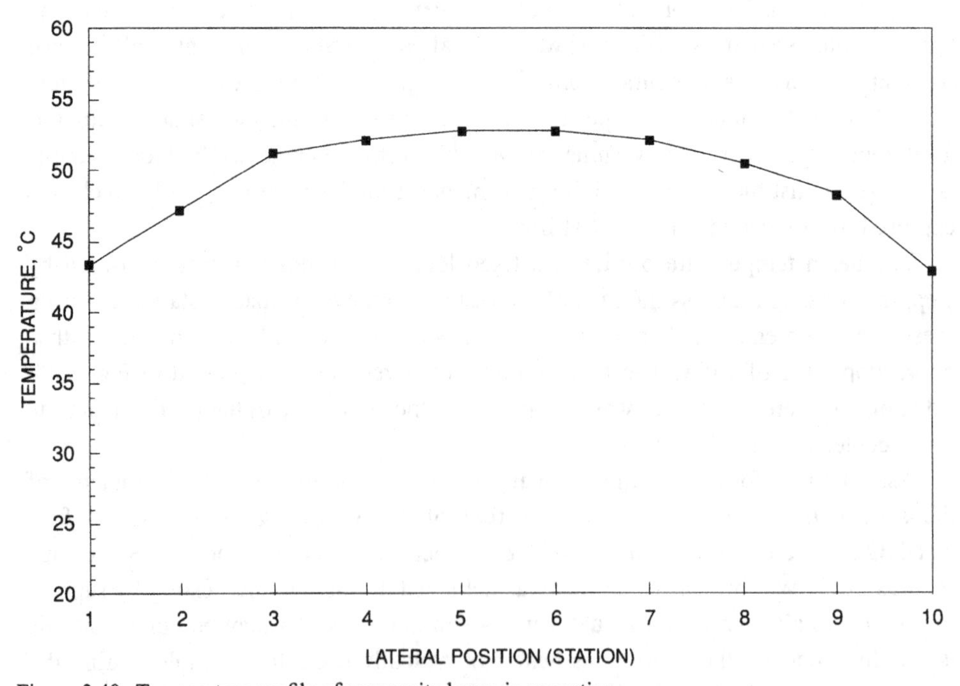

Figure 3.40. Temperature profile of composite beam in operation.

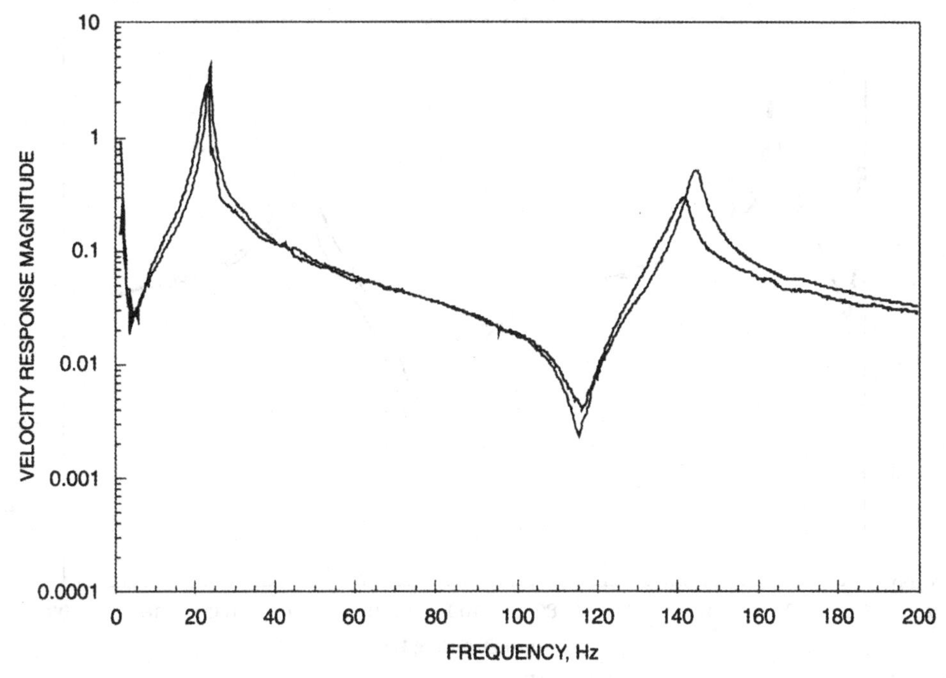

Figure 3.41. Comparison of Stage 2 and 3 FRFs for composite beam at 0.00 A.

of the matrix material had been exceeded. This would be a limitation in a practical system, but could be addressed to some extent by the design of the composite material and its curing process. No damage to the specimen was detected, and the material appeared to regain its original properties upon cooling to room temperature. That no permanent changes to the specimen occurred is borne out by the repeatability of the response data, demonstrated by the FRFs from Stages 2 and 3, which are compared in Fig. 3.41 for room temperature data and in Fig. 3.42 for a constant current of 0.50 A. The two curves plotted in each figure show data taken approximately 4 weeks apart, with the composite specimen having been dismounted, reinstalled, and extensively tested between the modal tests.

The experimental program has established that the unique properties of SMA can be exploited for use in this range of frequencies through a design strategy in which wires embedded near the surfaces are energized in a specific sequential manner, resulting in increased control bandwidth at the cost of reduced control authority compared to using all of the actuators at all times.

3.7 Applications of Shape Memory Alloys

The unique properties of SMA materials resulting from thermally induced phase transformation and superelastic behavior due to stress-induced phase transformation have been exploited commercially to produce a variety of products. The key factors that may determine the extent of applicability of SMA material are its transformation time and temperature. The time of transformation can be as fast as heat can be input and

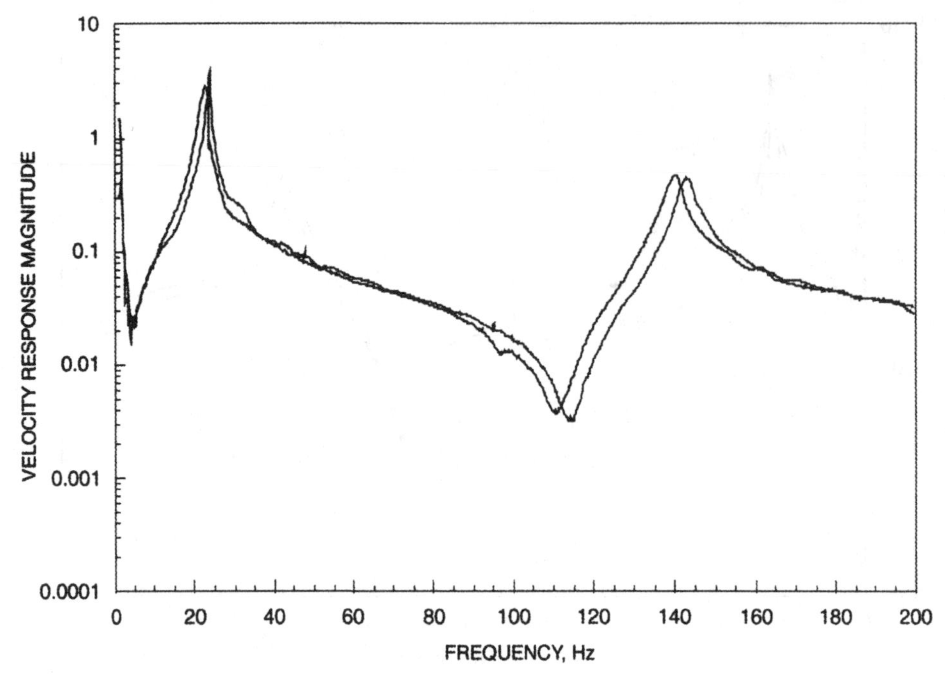

Figure 3.42. Comparison of Stage 2 and 3 FRFs for composite beam at 0.50 A.

removed, and is a function of the material's composition. The transformation temperatures can sometimes be tailored to suit the needs of a particular application through manipulation of constituents and processing. However, experts in the field warn us to be wary of trying to fit an SMA solution to every engineering situation. In this context the caution urged by (Hodgson, 1988) seems very relevant: "More than any other engineering material this author has encountered, the shape memory metal's properties are so interrelated with slight compositional variations, fabrication and processing history, training history... one must take each case on its own."

Nevertheless, popular applications include the following:

1. Shower springs that are activated above a certain temperature to shut off water that is too hot
2. Window latches that open and close windows automatically to protect plants in a nursery
3. Eyeglass frames that recover their original shape when accidentally bent out of shape
4. Blood clot filters that when inserted into the body open up at body temperature to arrest clots (Fig. 3.43)
5. Cryofit couplings that can be used to repair broken fuel or oil lines in an emergency (Fig. 3.44)
6. Circuit boards that disconnect from mother boards easily for repair or replacement (Fig. 3.45)
7. Braille characters that pop up when moved by SMA actuators and that can be "rewritten"

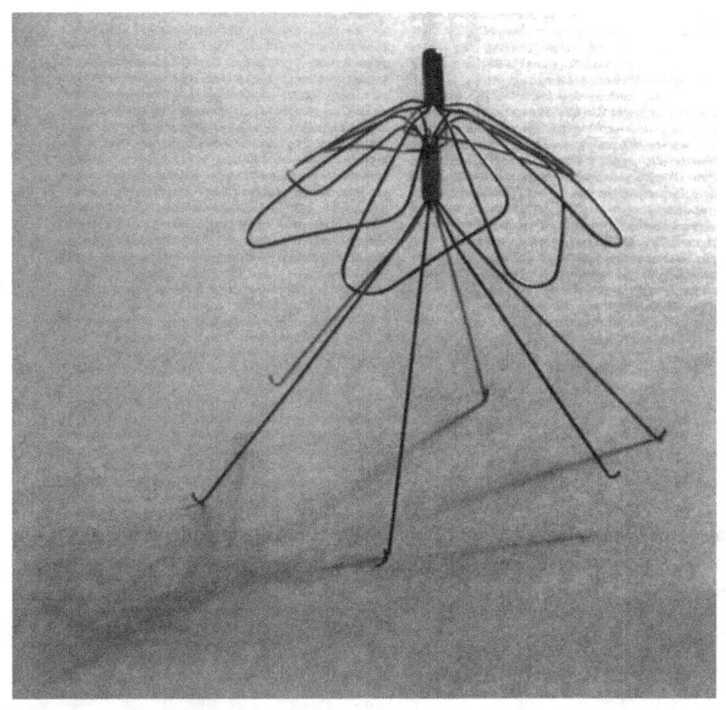

Figure 3.43. Filter to trap blood clots expands after insertion into a vessel. (Courtesy of Innovative Technologies International, Inc.)

8. Bracing wires to straighten teeth
9. NiTiNOL rods connected on to the spine to straighten it
10. Actuators for vanes controlling the flow of air through jet engines (Fig. 3.46)
11. Digital storage devices using thin-film SMA
12. Devices to control the sag of electrical power transmission lines

Several other applications are being considered and will undoubtedly attempt to find a market. Applications that exploit the inherent damping capacity of SMA are also under study.

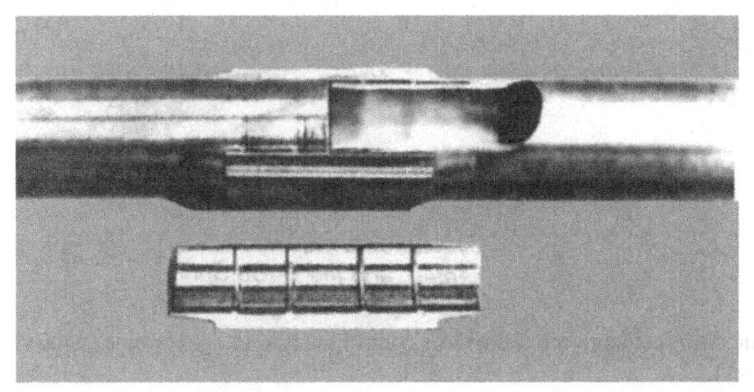

Figure 3.44. An SMA coupler contracts when heated to join pipes or tubes without welding. (Courtesy of David Hodgson, Shape Memory Applications, Inc., 1988.)

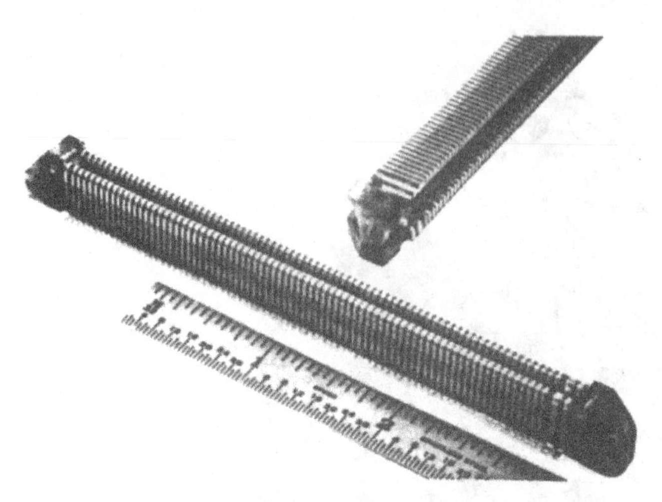

Figure 3.45. The clamping force of this card edge connector can be controlled to allow easy insertion and removal while ensuring firm contact during operation. (Courtesy of David Hodgson, Shape Memory Applications, Inc., 1988.)

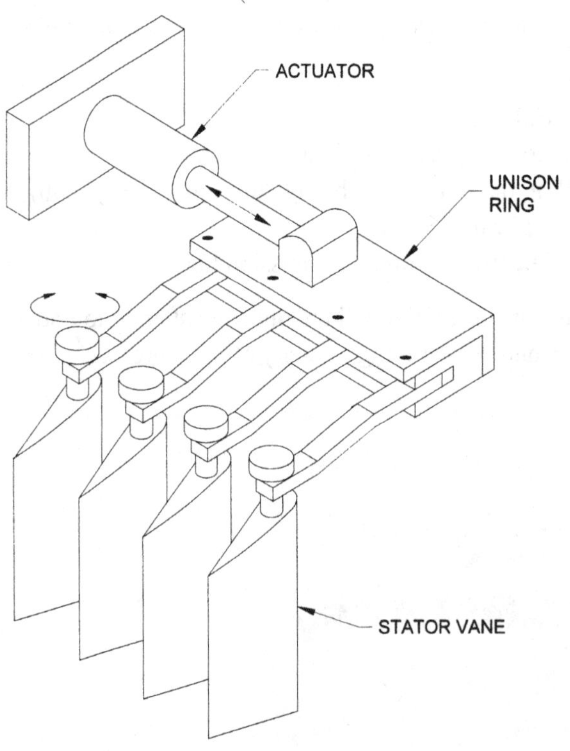

Figure 3.46. In this schematic, SMA actuators move the vanes controlling flow in a gas turbine engine (Srinivasan, unpublished notes).

The commonly used forms of the material come in wires of diameter 0.25 in. to less than 0.003 in. and strips as thin as 0.004 in. × 1 in. Stresses induced vary from 25 000 psi to 100 000 psi. Allowable strains vary from 8% for a single use to 3% for multiple uses. Fatigue life of a million cycles is not uncommon. An important impediment to far greater use and application of the material is its cost. Nickel-based alloys may cost more than $100/lb, although copper-based alloys are much cheaper, especially in large quantities.

3.8 Summary

Alloys that have the ability to change phase upon introduction of heat have found important applications in the medical field. The potential for use in engineering structures is limited by the range of temperatures within which phase change may take place and the rapidity with which these changes occur.

High force at "low" frequencies provided by shape memory elements can be used effectively in the design of actuators. Increased bandwidth of frequencies can be designed into smart structures by multiplexing actuator elements.

BIBLIOGRAPHY

Baz, A., K. Inman, and J. McCoy. 1990. Active vibration control of flexible beams using shape memory actuators. *Journal of Sound and Vibration* 140(3):437–465.

Boggs, R. N. 1993. How memory metals shape product designs. *Design News* 72–80.

Brinson, L. C. 1993. One-dimensional constitutive behavior of shape memory alloys: Thermomechanical derivation with non-constant material functions and redefined martensite internal variable. *Journal of Intelligent Material Systems and Structures* 4:229–242.

Buehler, W. J., J. V. Gilfrich, and R. C. Wiley. 1963. Effect of low-temperature phase changes on the mechanical properties of alloys near composition TiNi. *Journal of Applied Physics* 34:1475.

Ditman, J. B., L. A. Bergman, and T. C. Tsao. 1994. The design of extended bandwidth shape memory alloy actuators. In *Proceedings of the 35th structures, structural dynamics and materials conference*, number AIAA-94-1757. AIAA.

Gibson, R. F., and R. Plunkett. 1976. Dynamic mechanical behavior of fiber-reinforced composites: Measurement and analysis. *Journal of Composite Materials* 10:325–341.

Hansen, J. 1981. Metals that remember. *Science* 81:44–47.

Hodgson, D. 1988. *Using Shape Memory Alloys*. Shape Memory Applications, Inc. Sunnyvale, CA.

Inaudi, J. A., and R. Krumme. May 1995. On the reduction of sag of cables using nitinol devices. In *Proceedings: Shape memory alloys for power systems*, 14-1–14-19. Electric Power Research Institute.

Kauffman, G. B., and Isaac Mayo. 1993. The metal with a memory. *Invention & Technology* 9(2):18–23.

Liang, C., and C. A. Rogers. 1990. One-dimensional thermomechanical constitutive relations for shape memory materials. *Journal of Intelligent Material Systems and Structures* 1(2):207–234.

Liang, C., and C. A. Rogers. 1992. Design of shape memory alloy actuators. *Journal of Mechanical Design* 114:223–230.

O'Connor, L. 1995. Memory alloys remember two shapes. *Mechanical Engineering* 117(12):78–80.

Portlock, L. E., L. McD. Schetky, and B. M. Steinetz. 1995. Shape memory alloy adaptive control of gas turbine blade clearance. In *AIAA 1995 Joint Propulsion Conference*, number 95–2762.

Rogers, C. A., C. Liang, and J. Jia. 1989a. Behavior of shape memory alloy composite plates. Part I: Model formulations and control concepts. In *Proceedings of the 30th Structures, Structural Dynamics and Materials Conference*, number AIAA-89-1389-CP, 2011–2017. AIAA.

Rogers, C. A., C. Liang, and J. Jia. 1989b. Behavior of shape memory alloy composite plates. Part II. Results. In *Proceedings of the 30th Structures, Structural Dynamics and Materials Conference*, number AIAA-89-1331-CP, 1504–1513. AIAA.

Srinivasan, A. V. 1995. Galloping of transmission lines. In *Proceedings: Shape Memory Alloys for Power Systems*. Electric Power Research Institute.

Srinivasan, A. V., D. G. Cutts, and L. M. Schetky. 1991. Thermal and mechanical considerations in using shape memory alloys to control vibrations in flexible structures. *Metallurgical Transactions* 22A:623–627.

Srinivasan, A. V., D. M. McFarland, H. A. Canistraro, and E. K. Begg. 1997. Multiplexing of embedded NiTiNOL actuators for increased bandwidth in structural control. *Journal of Intelligent Material Systems and Structures* 8(3):202–214.

Waram, T. C. 1993. *Actuator Design Using Shape Memory Alloys*. 2d ed. San Anselmo, California: Mondotronics, Inc.

Wayman, C. M. 1992. Shape memory and related phenomena. *Progress in Materials Science* 36:203–224.

D. C. Wilson, J. R. Anderson, R. D. Rempt, and R. Ikegami. 1990. Shape memory alloys and fiber optics for flexible structure control. In *Proceedings of the SPIE: Fiber-Optic Smart Structures and Skins III*, number 1370, pages 286–295. SPIE.

PROBLEMS

1. In its martensitic state, a NiTiNOL wire is of diameter 0.5 mm and length 10 cm.

(a) The wire is given an initial strain of 2% at a temperature below M_f; its ends are then rigidly fixed. What force will be developed in the wire when it is heated to a temperature above A_f?

(b) The cycle of part (a) is then repeated but this time the wire is restrained by a spring of stiffness $k = 5\,000$ N/m. Find the displacement of the spring and the strain and force in the wire after heating.

2. An SMA wire 1 mm in diameter and 20 cm long is to be heated by an electrical current. The material of the wire is NiTi, and it is suspended in still air at an ambient temperature of 22 °C.

(a) Find the steady current I necessary to transform the wire, initially entirely martensitic, to austenite in 15 s. Also calculate the voltage that must be applied to produce this current and the electrical power dissipated.

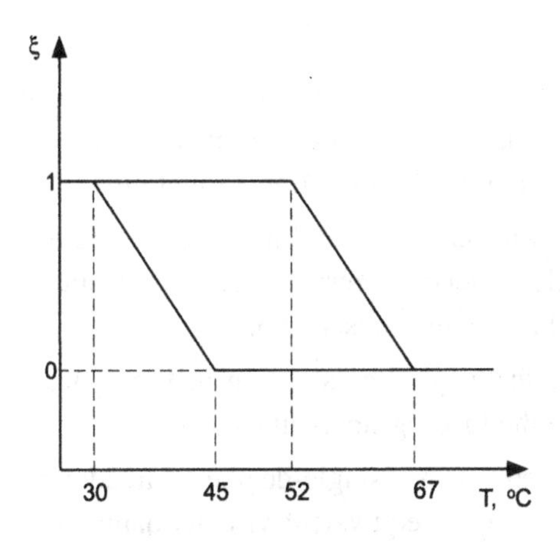

Figure 3.47. Linear model of SMA phase change (Problem 4).

(b) Calculate the time for the wire to return to with 5 °C of ambient temperature after the current is removed.

In both parts (a) and (b), neglect the heat of transformation of the material.

3. A strip of SMA is constrained at constant length in a tensile testing machine, subjected to a small initial load, and heated electrically. The strip is initially at 20 °C and the material transformation temperatures are $A_s = 40\,°\text{C}$ and $A_f = 80\,°\text{C}$. The load measured by the machine increases during the $M \rightarrow A$ transformation, then falls off somewhat with further heating.

(a) Explain the observed decrease in load.

(b) What would be the optimum temperature at which to maintain the SMA, i.e., that producing the maximum steady load?

4. A prismatic sample of SMA of size $1 \times 2 \times 10\,\text{mm}$ is initially martensitic, and recovers a 3% strain in the direction of its long dimension upon heating. The dependence of phase upon temperature is idealized as shown in Fig. 3.47.

(a) Assuming a uniform rate of heating, compute and plot
  (i) strain vs. temperature,

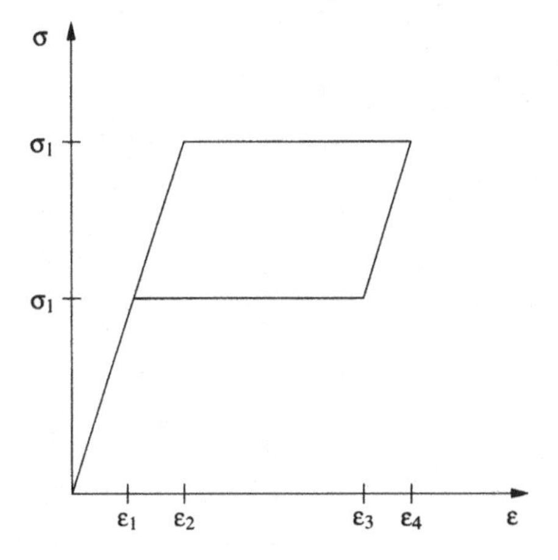

Figure 3.48. Idealized hysteresis loop from SMA wire test.

(ii) temperature vs. time, and

(iii) strain vs. time.

(b) Neglecting simple thermal strains, calculate the amount of mechanical work that could be done by the specimen during the $M \rightarrow A$ transformation.

5. A 15 cm length of 1-mm-diameter SMA wire is loaded and unloaded at room temperature. Because of the formation and recovery of stress-induced martensite, the stress-strain diagram from this test exhibits a hysteresis loop.

(a) Idealizing this response as elastic-perfectly plastic as shown in Fig. 3.48, calculate the energy dissipated during the loading-unloading cycle.

(b) If this wire is used as the damping element in a single-degree-of-freedom structure of stiffness k and mass m, find the equivalent viscous damping coefficient c.

4

Electrorheological and Magnetorheological Fluids

4.1 Introduction

In this chapter we consider fluids whose properties change in response to an applied electric or magnetic field. These are known as *electrorheological* (ER) or *magnetorheological* (MR) fluids, respectively, because the most remarkable field-induced change is a tremendous increase in their ability to support shear stress. Most engineering applications of ER and MR fluids exploit their controllable yield stress to vary the coupling or load transfer between moving parts, for example in dampers and clutches.

Although application details differ because of the requirements of generating strong electric or magnetic fields, the basic physics describing how ER and MR fluids' material properties change, and the design of mechanical devices to capitalize on these changes, are similar for the two types of controllable fluid. Therefore, we shall treat them simultaneously, distinguishing between the ER and MR responses as necessary but in general being more concerned with common features than with differences. We shall first examine the phenomenology and modeling of ER and MR fluids, then review some of their current and emerging applications.

4.2 Mechanisms and Properties

The ER and MR effects are the result of the formation of structures within a fluid in response to an electric or magnetic field. These structures, actually aggregations of solid particles, dominate the flow of the fluid, and can prevent flow entirely at lower stresses. In this section, we discuss these microscopic phenomena and present basic models of the corresponding macroscale fluid mechanics.

4.2.1 Fluid Composition and Behavior

Both electrorheological and magnetorheological fluids are suspensions of particles in inert carrier liquids. The particles, typically of the order of 1 to 10 μm in size, are

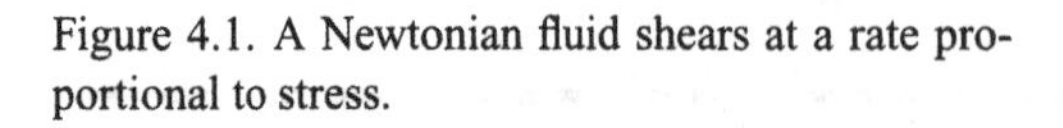

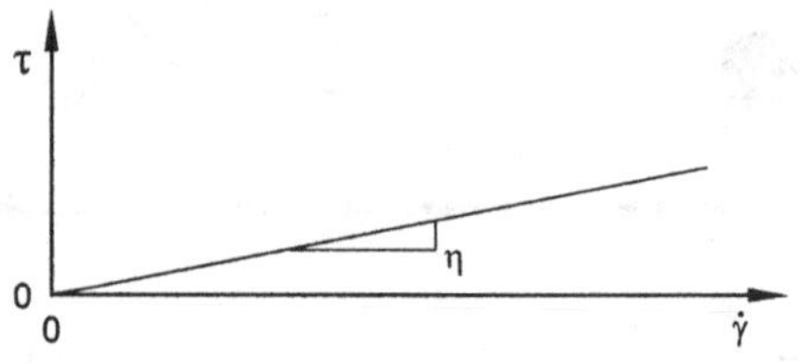

Figure 4.1. A Newtonian fluid shears at a rate proportional to stress.

added to fluids, such as mineral oils or silicone oils, in weight fractions as large as 50%, with fractions of around 30 wt% being common. Most ER and MR fluids also contain small amounts of additives that affect the polarization of the particles or stabilize the structure of the suspension against settling, but for many engineering purposes these may be neglected in modeling the fluids' mechanical response.

In the absence of an external electric or magnetic field, an ER or MR fluid may be characterized as Newtonian, i.e., as resisting shear strain γ with a shear stress τ proportional to the product of the strain rate $\dot{\gamma}$ and viscosity η:

$$\tau = \eta\dot{\gamma}. \tag{4.1}$$

This response is represented by the line passing through the origin in Fig. 4.1, which shows shear stress as a function of strain rate. This is widely acknowledged as an approximation – most ER and MR fluids are non-Newtonian even when no field is applied because of their heavy loading of solid particles and, to some extent, because of the additives they contain. However, in most applications the field-induced component of the shear stress is much larger than the $\eta\dot{\gamma}$ term and eq. (4.1) is an adequate model of the rate-dependent part of the total shear stress.

In the zero-field or "off" state the particles are usually assumed to be uniformly distributed. In some applications the normally occurring flow of the fluid is sufficient to prevent settling. In others, gentle stirring may be required to redistribute the particles, or chemical additives may be introduced to stabilize the particle distribution and combat sedimentation. As will be seen below, it is generally advantageous that the zero-field viscosity η be as small as possible, but measures that increase viscosity or produce a thixotropic mixture may be required in some applications. For purposes of analysis and design, we will assume a uniform distribution of particles until an external field is applied to the fluid.

The effect of an electric field E on an ER fluid, or of a magnetic field H on an MR fluid, is to cause the particles to form chains, or *fibrils*, in the direction of the field. This process of fibration occurs in a few milliseconds after application of the field. When there is no motion of the fluid or of the walls of its container, the fibrils are static structures and span the gap between the walls if the particle fraction is large enough. (This is one reason fluids with low particle fractions exhibit weak ER and MR effects.) In an MR fluid, the formation of particle chains occurs when the magnetically polarizable particles move into alignment with the applied field and are then drawn together like magnets whose opposite poles attract the adjacent particles in the chain. The formation of fibrils in an ER fluid in response to an external electric field happens in a similar way, but the chemistry underlying the electrical polarization of the particles has been the subject of much research attention. It now appears that ER fluids are divisible into two classes depending upon the mechanism by which particle polarization and interaction occur. One type of ER fluid requires the presence of some amount of water in order to manifest an electrorheological response; other, anhydrous ER fluids contain no water, and particle chain formation is thought to occur in them by a different mechanism (Weiss, Carlson, and Coulter, 1993a). For our purposes, it is unimportant to distinguish which electrical polarization mechanism is at work in a given ER fluid, but we note in passing that this contributes to the sensitivity of ER fluids to water contamination.

Magnetorheological fluids differ from conventional "magnetic fluids," which contain particles of much smaller size, typically of the order of 10 nm. The effect of Brownian motion is greater at this scale, and prevents the particles from forming fibrils in the presence of a magnetic field. The magnetic fluid instead experiences a body force proportional to the magnetic field gradient, and may flow in response to this force. This behavior is exploited, for example, in sealing applications; however, here we are concerned only with the MR and ER effects exhibited by fluids containing larger particles.

When an ER or MR fluid flows, or when there is relative motion between the walls of its container, shear strain occurs in the fluid and a shear stress distribution develops across the fluid. To the extent the Newtonian model described above holds, this stress distribution can be calculated by using the viscous flow equations of elementary fluid mechanics. When a field oriented normal to the direction of flow or motion is applied, fibrils form across the flow, and because of the motion of the fluid or the walls these fibrils are broken and must reform. The continual breaking and reforming of these particle chains results in a force resisting the motion of the fluid or walls and gives rise to the field-dependent component of the shear stress τ. In most cases, this component is much larger than the viscous shear stress $\eta\dot{\gamma}$. It is this large, controllable shear stress that makes these fluids useful in mechanical systems.

The ER or MR shear stress increases with increasing field strength, and is typically proportional to the field strength raised to a power between 1 and 2. In some configurations this stress component should in theory vary with the square of the field, but experimental data more commonly vary less rapidly, in some cases being almost linear in the field strength. The upper limit on the induced shear stress occurs when an MR fluid reaches magnetic saturation or when an ER fluid breaks down electrically, typically at fields strengths of around 250 A/mm or 4 kV/mm, respectively.

4.2.2 Discovery and Early Developments

The earliest reports of the ER and MR effects, discussed below, appeared in the late 1940s, and are remarkable for their insight into the physical phenomena, for the scope and thoroughness of the test programs that had been carried out, and for the degree to which their authors anticipated present-day applications and the problems they would present. Following the appearance of those initial papers, development of ER fluids and applications proceeded sporadically, while MR fluid technology attracted little interest. Steady work with ER fluids began in the middle to late 1980s, after which research on MR fluids attracted greater attention. Today, the two technologies are roughly equally mature, and research continues on both the fluids themselves and their applications.

Rabinow (1948) reported on an experimental program undertaken at the U.S. National Bureau of Standards for the Army's Chief of Ordnance. The latter anticipated a need for high-speed mechanical input/output devices for use with the electronic computing equipment then beginning to appear. Rabinow described several potential applications for the MR fluid studied, consisting of "finely divided iron" (in one case the particle size is given as 8 μm) in oil, but focused on clutches of various designs and sizes. With regard to this application, he made a number of observations borne out by more recent work and discussed below. In addition to results obtained with a number of small devices and test rigs, he presented construction details and test data for a clutch housed in a cylinder of approximately 15 cm diameter and 17 cm length that was capable of transmitting torques of up 130 N m.

At almost the same time, Winslow (1949) published his account of a lengthy program of research into the properties and applications of an ER fluid consisting of semiconductive solid particles of about 1 μm diameter suspended in oil. As well as describing and demonstrating the physical phenomena of fibration, this paper presents the results of rheological experiments and fatigue tests. Prototype devices built and tested included a valve with no moving parts, a clutch or brake, and even a loudspeaker driven by a small ER clutch. Although Winslow's interest was chiefly in ER rather than MR fluid technology, he did briefly discuss the MR effect and even mentioned a suspension of magnetic iron oxides in oil that could respond to both electric and magnetic fields, separately or simultaneously.

Both of these early works are notable not only for the scope of the research programs they describe and the amount of basic science they present, but also for their practical observations. For example, Rabinow wrote about a permanent-magnet "lock" or seal to prevent contamination of bearings located near a chamber containing MR fluid, and Winslow mentioned possible gains in the efficiency of a screw-type pump. Perhaps because the field is still young, in the sense that fundamental work remains to be done, this cognizance of the demands of practical applications is a feature often found in more recent research papers as well.

4.2.2.1 The Bingham Plastic and Related Models

One might expect that the formation of fibrils within an ER or MR fluid would increase the fluid's viscosity, but in fact the slope of the shear stress versus shear strain rate curve, and thus the viscosity, changes little if at all. The effect of the fibrils is instead to produce a shear stress that is largely independent of the strain rate; this is commonly

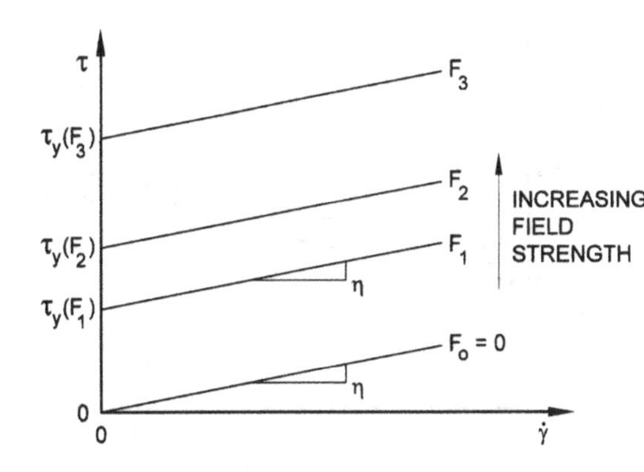

Figure 4.2. Shear stress versus shear strain rate for the Bingham plastic material model.

referred to as the *yield stress* and denoted τ_y. Adding this term to the Newtonian model of Section 4.2.1 results in the Bingham plastic model, which has the stress-strain rate relation

$$\tau = \tau_y(F) + \eta\dot{\gamma}, \tag{4.2}$$

where in a given application F is the strength of the applied electric or magnetic field (i.e., E or H). The response predicted by this model is plotted in Fig. 4.2, which depicts the strong dependence of the yield stress on the field strength. This model, or extensions of it that predict similar overall response, is by far the most popular for use in the design of devices that depend on the post-yield shear resistance of an ER or MR fluid. Furthermore, it is nearly ubiquitous in the literature, and most ER and MR fluids are characterized in part by their zero-field viscosity and field-dependent yield stress.

In practice, the dynamic viscosity (the slope of the τ vs. $\dot{\gamma}$ curve) is determined by a linear regression fit of a line to experimental data, and the intersection of this line with the shear stress axis is taken as the value of the yield stress τ_y. Although this is a good approximation at higher strain rates and is entirely adequate for most dynamic response calculations, the data measured at small strain rates can depart from this idealization. Initiating motion or flow requires overcoming a static yield stress $\tau_{y,s}$, which is often larger than the dynamic yield stress $\tau_{y,d}$, but the measured stress quickly falls to its dynamic value as shown by the dashed path in Fig. 4.3. Once τ has reached its dynamic value, it tends to follow the fitted straight line towards τ_y as $\dot{\gamma}$ decreases. Reasons for this behavior may include the reattachment to the walls of the field-induced fibrils, which are broken near their ends by the bulk shear of the fluid accompanying flow. It has also been observed that $\tau_{y,s}$ can sometimes rise significantly after long periods of no fluid motion, suggesting an additional, slower process can take place. Finally, there is the possibility of simple stiction due to friction between moving parts, seals, etc., especially in equipment built for use outside the laboratory. Whatever its source, this deviation from ideal behavior is often small enough that it may be ignored, especially in applications where transient response is unimportant.

More detailed models than the Bingham plastic of eq. (4.2) can of course be devised and fitted to experimental data. It may be necessary, for instance, to capture the dynamics associated with the momentum of the fluid, or to represent the finite compliance of the

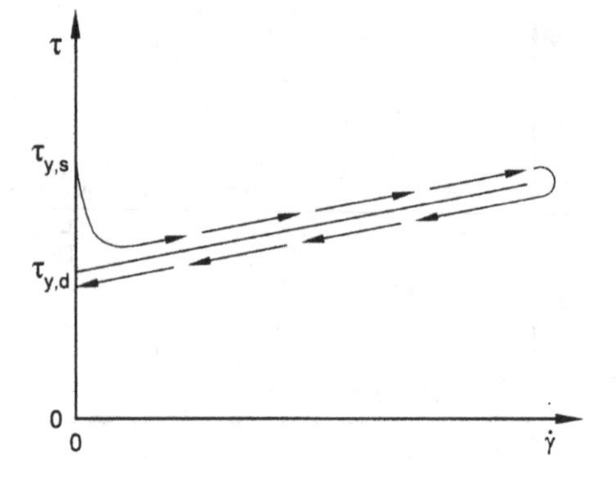

Figure 4.3. Static and dynamic shear stress responses.

container. An example of this extension of the Bingham model in the case of an MR fluid damper is presented below in Section 4.3.2.

4.2.2.2 Pre-Yield Response

According to the Bingham plastic model, stress less than the yield stress τ_y produces no flow of the ER or MR fluid; but in reality the fluid naturally responds to stress in this range, and for many purposes it may be regarded as a viscoelastic solid. Figure 4.4 shows typical stress-strain characteristics for an ER or MR fluid loaded up to and beyond yield. Note that yield occurs at approximately the same strain γ_y regardless of field strength, while τ_y increases with F as discussed above. For clarity we have shown the yield strain as corresponding to the peak stress on each curve, but in practice the correct definition of yield for these materials is not so clear. Weiss, Carlson, and Nixon (1994) suggest that the onset of nonlinearity in the storage modulus (defined below) is an accurate indicator of yield, and that it correlates well with measured static yield stresses.

The shear stiffness of viscoelastic solids, including unyielded ER and MR fluids, is often represented by the complex shear modulus $G^* = G' + jG''$. The real part G' is called the *storage modulus* and measures the material's ability to elastically store strain energy, while the imaginary part G'' is termed the *loss modulus* and is associated with the dissipation of energy during deformation. The *loss factor* is then defined as the ratio of the loss modulus to the storage modulus, G''/G', and can be determined

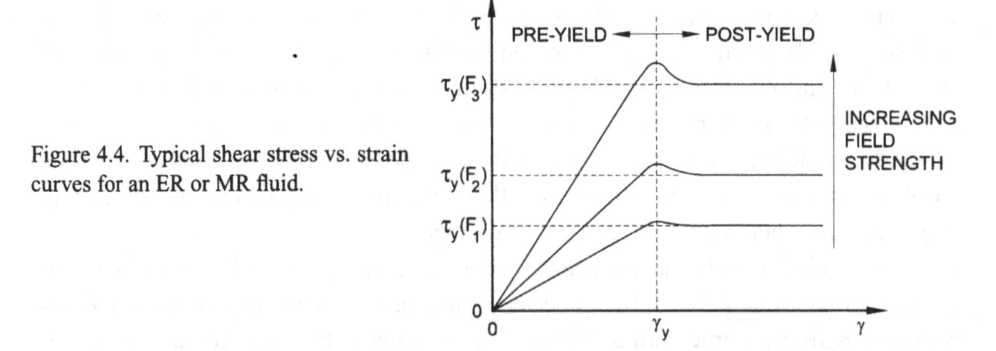

Figure 4.4. Typical shear stress vs. strain curves for an ER or MR fluid.

by measuring the phase difference between a strain wave input to a material and the resulting stress wave (Weiss et al., 1994). A predominantly elastic material will exhibit a small phase difference and a very small loss factor, typically less than 0.1. In a viscous material the phase difference will approach 90° and the corresponding loss factor will be quite large.

4.2.2.3 Post-Yield Flow and Device Geometry

In many devices where ER or MR fluids are employed, at any time a small portion of the fluid is subject to the applied electric or magnetic field, while the remainder is free to flow as a conventional, low-viscosity fluid. Ordinarily the field is created across a small gap whose surfaces serve as both electrodes (in the case of an ER fluid) or pole pieces (in the case of an MR fluid) and as the walls of a channel confining the fluid. For purposes of analysis, these walls are often modeled as parallel, flat plates; this is the only configuration we shall consider in detail here. The resulting equations are frequently applied to annular or other nonflat geometries with acceptable results, but more realistic models are available, e.g., for cylindrical dampers (Spencer et al., 1998).

Restricting consideration to a gap formed by two parallel flat plates, we must still take account of how shear is created in the fluid. The two possibilities are illustrated in Fig. 4.5, where it is shown that shear may be produced in the fluid either by forcing it through the gap under pressure (fixed-plate configuration) or by moving one plate with respect to the other (sliding-plate configuration). In either case, and for either type of fluid, the gap is small in the direction of the field, often well under 1 mm, and of the order of millimeters in length in the direction of flow or motion. This close spacing of the plates is required in order to produce a field strong enough to activate the fluid; the ER and MR effects, that is, the formation of fibrils in the direction of the applied field, will occur over a greater distance within the fluid if an adequate field can be generated.

The analysis of ER or MR fluid flow or plate motion can be carried out by a straightforward extension of elementary fluid mechanics. We present here a partial summary of results from Coulter, Weiss, and Carlson (1993); a detailed derivation may be found in that paper. Although the paper cited treats only the case of an ER fluid, most of the results are readily applicable to MR fluids used in similar geometries.

Consider first the pressure-driven flow of ER fluid between fixed electrodes, sometimes called valve-mode flow. A simple analysis based on a differential control volume shows how the pressure gradient Δp produces a shear stress distribution through the fluid layer. This distribution is found to be of the form

$$\tau(z) = -p'\left(z - \frac{h}{2}\right), \tag{4.3}$$

where

$$p' = -\frac{dp}{dx} = \text{constant} \tag{4.4}$$

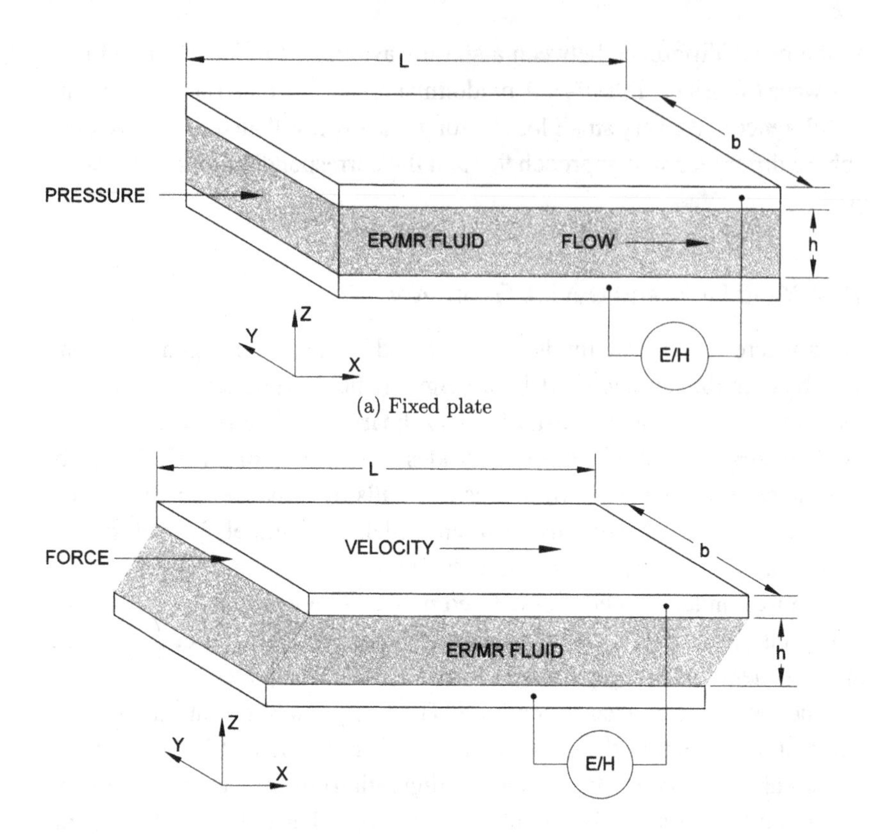

Figure 4.5. Gap geometries and fluid shear modes.

(note the choice of sign in the definition of p'). Although this distribution is independent of the nature of the material in the gap, if that material exhibits a yield stress then a critical pressure is required to cause flow; this critical pressure is related to the gap geometry and fluid yield stress by

$$\left|\left(\frac{dp}{dx}\right)_c\right| = \frac{2\tau_y}{h}. \tag{4.5}$$

If the pressure exceeds this critical level, flow will occur, and this flow will be characterized by shear of a layer near each wall while the central portion of the fluid is unyielded (resembling plug flow), as shown in Fig. 4.6. Denoting by h_1 the thickness of one of these sheared layers, the equation for the axial velocity distribution can be written

$$\frac{du}{dz} = \frac{p'(h_1 - z)}{\eta}. \tag{4.6}$$

The volumetric flow rate Q can be obtained by integration of this distribution,

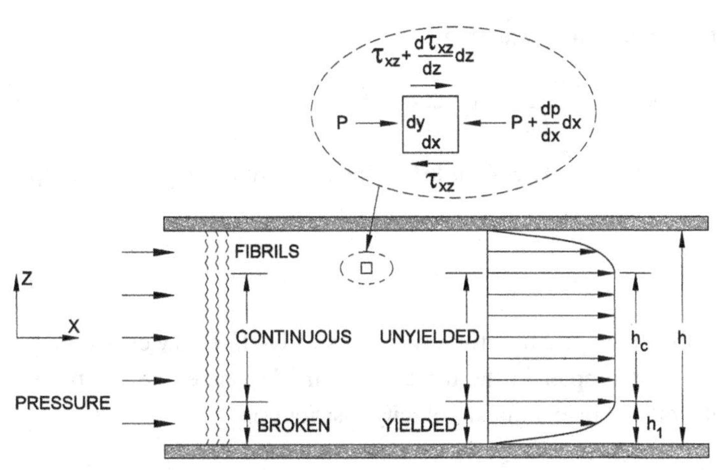

Figure 4.6. Flow velocity distribution through the thickness of a yielded ER or MR fluid.

resulting in

$$\frac{Q}{b} = \frac{(p'h - 2\tau_y)^2(p'h + \tau_y)}{12p'^2\eta}. \tag{4.7}$$

This relation among flow rate, pressure drop, and yield stress can be put into the nondimensional form

$$P^{*3} - (1 + 3T^*)P^{*2} + 4T^{*3} = 0, \tag{4.8}$$

where

$$P^* = \frac{bh^3 p'}{12Q\eta}, \qquad T^* = \frac{bh^2\tau_y}{12Q\eta}. \tag{4.9}$$

Two approximations to this equation are commonly made, depending upon the flow regime of interest. When $T^* < 0.5$, the equation

$$P^* = 1 + 3T^* \tag{4.10}$$

is accurate within 5%. Returning to the use of dimensional quantities, this expression is equivalent to

$$\Delta P = \Delta P_{0,HF} + \Delta P_{ER,HF} = \frac{12\eta QL}{bh^3} + 3\frac{L}{h}\tau_y \tag{4.11}$$

where P_0 is the viscous component of the pressure drop, P_{ER} is the contribution of the field-induced shear stress, and the subscript *HF* indicates that this approximation is applicable under conditions of high flow. Similarly, for low-flow conditions corresponding to $T^* > 200$, a suitable approximation can be shown to be

$$P^* = \frac{2}{3} + 2T^*, \tag{4.12}$$

which is equivalent to the dimensional equation

$$\Delta P = \Delta P_{0,LF} + \Delta P_{ER,LF} = \frac{8\eta QL}{bh^3} + 2\frac{L}{h}\tau_y. \tag{4.13}$$

Comparing these limiting cases, we may conclude that the ER term in the pressure drop is of the form

$$\Delta P_{ER} = C\frac{L}{h}\tau_y, \tag{4.14}$$

where C is a constant in the range of 2 to 3 for any given steady-state flow conditions.

The situation is somewhat simpler in the direct-shear mode where one electrode moves with respect to the other, since a linear velocity distribution

$$\dot{\gamma} = \frac{S}{h} \tag{4.15}$$

may be assumed through the fluid thickness. Combining this with the simple expression

$$f = \tau A \tag{4.16}$$

relating the resultant force f to the plate area A leads to

$$f = \frac{SLb\eta}{h} + Lb\tau_y, \tag{4.17}$$

and we may again identify viscous and shear-stress terms.

In the same paper, Coulter et al. (1993) examine the significance of the *control ratio*, a measure of the effect of the ER fluid on the mechanical performance of the system of which it is a part. The control ratio is defined for a fixed-plate device as the pressure drop when the field is on to the pressure drop under zero-field conditions. From this definition, the minimum amount of ER fluid needed to achieve a desired degree of control is found to be

$$V_F = \frac{12}{C^2}\frac{\eta}{\tau_y^2}Q\Delta P_{ER}\left(\frac{\Delta P_{ER}}{\Delta P_{0,HF}}\right), \tag{4.18}$$

where the subscript F denotes the fixed-electrode configuration. The power required to maintain the electric field, and thus the electrorheological yield stress τ_y, is then

$$W_F = \frac{12}{C^2}\frac{\eta}{\tau_y^2}Q\Delta P_{ER}\left(\frac{\Delta P_{ER}}{\Delta P_{0,HF}}\right)EJ \tag{4.19}$$

where J is the current density (current per unit electrode area) and, as before, E is the electric field strength. The same sort of calculations for the sliding-electrode geometry, for which the control ratio is defined as the ratio of the resultant force with the field on to that with it off, lead to a minimum fluid volume of

$$V_S = \frac{\eta}{\tau_y^2}Sf_{ER}\left(\frac{f_{ER}}{f_0}\right) \tag{4.20}$$

and power requirement

$$W_S = \frac{\eta}{\tau_y^2} S f_{ER} \left(\frac{f_{ER}}{f_0} \right) EJ. \tag{4.21}$$

It is generally desirable to minimize both fluid volume and power consumption, and thus the value of a high control ratio is immediately apparent. Therefore, we again conclude that low zero-field viscosity and high field-induced yield stress are advantageous. Accordingly, we would wish to minimize the quantity η/τ_y^2, which, because of its role in design equations like those above, is sometimes used as a figure of merit in comparing ER or MR fluids.

4.2.2.4 Other Effects

In addition to the primary ER or MR effect, the existence of the field-controllable yield stress τ_y, magneto- and electrorheological fluids exhibit a number of other, generally weaker responses to changes in their environment. Several of the references cited at the end of this chapter mention some of these, often in the context of a particular application. Brooks (1993) discusses ER fluid behavior in detail, and we summarize some of the results from that paper in this section. Once again, similar findings might be expected for a typical MR fluid, with the exception of those pertaining to electrical current and current density.

The parameters of greatest interest in many ER fluid applications are the yield stress τ_y (sometimes called the "excess shear stress" because of the manner in which it adds to the viscous shear stress term), the viscosity η, the particle volume fraction φ, and the electrical current I that flows through the fluid. Assuming constant electrode area, we may work with either current I or current density J; likewise, for constant electric field strength E, power and current are linearly related.

Temperature variations affect an ER fluid predominately through the inert liquid phase of the suspension, that is, the suspended particles are relatively little changed. The carrier fluid may expand or contract with temperature changes, altering the effective volume fraction of the particles, and its viscosity can vary with temperature. Electrical current tends to increase with increasing temperature, with the result that the steady-state current through the ER fluid depends in part upon the net rate of heat transfer from the ER device. On the other hand, the current tends to decrease with increasing shear rate $\dot{\gamma}$. Brooks (1993) observes, "The current density is a strong function of both the field strength and the fluid temperature, and a weak function of both the shear rate and the volume fraction" (p. 412).

To study the influence of particle volume fraction, the relative viscosity is defined as

$$\eta_r = \frac{\eta}{\eta_f} \tag{4.22}$$

where η_f is the viscosity of the carrier fluid, while η is the viscosity of the ER suspension, taken to depend upon φ. In general, increasing particle volume fraction results in increasing suspension viscosity, although the increase is slow for φ less than about 25%. Brooks suggests this dependence is of the form

$$\eta_r = \left(1 - \tfrac{1}{2}k\varphi\right)^{-2} \tag{4.23}$$

where $k = 2/\varphi_m$, with φ_m being the maximum packing fraction of the particles. From geometric considerations, the numerical value of φ_m can be found to range from 0.74 if the particles are arranged in a face-centered cubic or close-packed configuration, to 0.525 if their packing is simple cubic. Fitting this model to experimental data produced a value of $k = 3.661$, corresponding to $\varphi_m = 0.55$. This agrees well with the experimentally determined packing fraction of $\varphi_m = 0.58$. On the basis of these values, the viscosity of an ER fluid may be predicted to be

$$\eta = (1 - 1.83\varphi)^{-2}\eta_f. \tag{4.24}$$

Particle volume fraction also affects the yield stress (although less rapidly than it does viscosity). For the data of the cited paper, good correlation is found with the model

$$\tau_y = 43.20\varphi^{2.5}. \tag{4.25}$$

4.2.3 Summary of Material Properties

Although testing procedures and even the definitions of material properties of ER and MR fluids are not as thoroughly standardized as in longer-established areas of materials science, a substantial body of data exists in the literature of the field. Some typical values are presented here so that the reader may gain an appreciation of the approximate capabilities of a generic ER or MR fluid, but we must stress that in a particular application much better performance may be achieved with the selection of a special purpose fluid. Some vendors have engaged in extensive research programs, which are bearing fruit in the form of commercially available fluids with broad operating temperature ranges, low toxicity, and other desirable characteristics. Further data, and an extensive tabulation of fluid compositions (much of it drawn from patent literature), can be found in Weiss et al. (1993a), which also addresses the problem of comparing experimental fluid data from different sources.

Both ER and MR fluids have similar zero-field viscosities, of the order of 0.1 to 1.0 Pa s, but they differ markedly in the maximum yield stress they can support. Typical maximum values of τ_y for ER fluids are in the range of 2 to 5 kPa, while MR fluids are often an order of magnitude stronger with maximum τ_y in the range of 50 to 100 kPa. Their densities are similar, 1 to 2 g/cm^3 for ER fluids and 3 to 4 g/cm^3 for MR fluids. (The foregoing values are taken from Table 3 of Spencer and Sain, 1997.) Current densities observed in ER devices are of the order of 10^{-6} to 10^{-3} A/cm^2 (Weiss et al., 1993a).

Duclos (1988) suggests that typical ER fluid parameters are

$$\tau_y = 2\,\text{kPa}, \tag{4.26}$$

$$\eta = 100\,\text{mPa s}, \tag{4.27}$$

$$E = 4\,\text{kV/mm}, \tag{4.28}$$

$$J = 0.1\,\mu\text{A/cm}^2 \tag{4.29}$$

at room temperature. Bearing in mind that a decade of research has produced improvements in ER fluids since this paper appeared, these values may suffice for back of the envelope calculations, and for the exercises at the end of this chapter.

In many designs it will be desirable to compare the potential performance of ER and MR fluids. This can be done in part by using the figure of merit relating zero-field viscosity and the square of the field-induced yield stress, introduced in Section 4.2.2.3. Spencer and Sain (1997) give values of η/τ_y^2 in the ranges 10^{-10} to 10^{-11} s/Pa for MR fluids and 10^{-7} to 10^{-8} s/Pa for ER fluids, while Weiss et al. (1993b) uses values near the low ends of these ranges for both types of fluid. Recall that, in general, the smaller the value of this ratio the better, although we will see below that MR fluids do not enjoy quite the overwhelming advantage these numbers might suggest.

Finally, we should note that ER fluids are more sensitive to contaminants, including water, than are their MR counterparts. This is mostly due to the effects of water on the electrochemical mechanisms of particle polarization and fibril formation, and can pose difficult quality control problems in manufacturing and in applications. MR fluids are typically much more tolerant of most impurities (Carlson, Catanzarite, and St. Clair, 1996).

4.3 Applications of ER and MR Fluids

The majority of engineering applications of ER and MR fluids, and those that have received by far the largest share of attention in the relevant literature, exploit the fluids' controllable yield stress τ_y. We review below a few of the publications pertaining to two basic types of devices: clutches and dampers. The latter especially have many potential applications in smart structures, where they are often used in semi-active control schemes to reduce the levels of vibration caused by live loads, strong winds, or earthquake ground motions. Following our discussion of these devices, we shall briefly mention some other novel applications of ER and MR fluids.

For control and other applications, it is important to understand what is meant by a semi-active device. Simply put, a *semi-active* system is one whose properties change in response to a command signal, offering the possibility of exerting a large control effort with little energy input; this is usefully compared to a *passive* system, whose characteristics are fixed, and to an *active* system, which acts on its environment using energy from another source. Examples of each are a simple viscous dashpot with fixed damping coefficient (a passive system), a semi-active shock absorber with a variable orifice that allows its damping coefficient to be controlled (a semi-active system), and an electromechanical actuator arranged so as to generate a force in response to the feedback of a velocity signal (an active system). The semi-active damper offers the possibility of real-time tuning of its properties without the costs, including energy input, of the fully active damper. It is commonly found that an optimal semi-active design can deliver most of the performance of a fully active design at a savings of both complexity and power consumption.

Before delving into specific applications, we can make some general observations about the implications of material properties for the choice of an ER or MR fluid. In some cases, the much greater yield stress achievable with an MR fluid may be the overriding consideration. In others, the high voltage required to create the electric field needed to energize an ER field may be unacceptable from the standpoint of safety, or might dictate the use of uncommon or nonstandard connectors or other parts. The power supplies used with ER fluid devices typically must provide a few mA at several thousand volts, while those used with MR devices are low-voltage (12–24 V) supplies capable of providing a few amperes of current. Generally speaking, the high-voltage power supply will be more expensive. The magnetic circuit that produces the field H across the fluid-filled gap in an MR device usually contains a significant quantity of steel, the weight of which may be important in some applications; on the other hand, the total weight of comparable ER and MR systems may be roughly the same, or the MR system may even be lighter (Weiss et al., 1993b).

On the basis of the equations developed in Section 4.2.2.3, we can conclude that for the same control ratio, force, and speed, the ratio of the volume of ER fluid that would be required to that of MR fluid is (Carlson et al., 1996)

$$\frac{V_{ER}}{V_{MR}} = \frac{\eta_{ER}/\tau_{ER}^2}{\eta_{MR}/\tau_{MR}^2}, \tag{4.30}$$

and because the zero-field viscosities of ER and MR fluids are approximately the same,

$$\frac{V_{ER}}{V_{MR}} \approx \left(\frac{\tau_{MR}}{\tau_{ER}}\right)^2. \tag{4.31}$$

This ratio is typically of the order of 100 to 1 000, implying that the volume of the fluid gap in an MR device can often be much smaller than that of an ER device operating under the same conditions. To compare the energy W in the field throughout this volume for each type of fluid, we compute

$$W_{ER} = V_{ER}\left(\tfrac{1}{2}\kappa\varepsilon_0 E^2\right), \tag{4.32}$$

where κ is the relative dielectric constant of the ER fluid, ε_0 the permitivity of vacuum, and E the applied electric field, and compare this to

$$W_{MR} = V_{MR}\left(\tfrac{1}{2}BH\right), \tag{4.33}$$

where B is the magnetic induction of the MR fluid in the gap and H the applied magnetic field. Taking the ratio of these expressions gives us

$$\frac{W_{ER}}{W_{MR}} = \left(\frac{V_{ER}}{V_{MR}}\right) \frac{\kappa \varepsilon_0 E^2}{BH}. \tag{4.34}$$

For a volume ratio of 100 and using typical values of the other parameters,

$$\frac{W_{ER}}{W_{MR}} \approx 1, \tag{4.35}$$

and therefore the external power required will be approximately equal in designs of similar capabilities whether ER or MR fluid is used. In addition to this estimate based on the ability of the fluid to store electromagnetic energy, one needs consider electrical losses resulting from dissipation in the ER fluid itself or in the coil of the electromagnet used to create the field H. If the ER fluid is energized by an alternating-current field, the circuitry must be adequate to supply not only the current that flows through the fluid but any displacement current associated with the polarized particles' response to the reversal of the applied field (Weiss et al., 1993a).

4.3.1 Clutches

A natural application of the controllable yield stress of ER and MR fluids is in clutches used to transfer torque between rotating, nominally rigid mechanical components. The fluids' ability to withstand shear deformation without suffering damage, coupled with their fast response time and the potential to smoothly control the coupling between the input and output shafts, offer advantages that were recognized early on (both Rabinow and Winslow included clutches among their prototype devices).

Most ER and MR fluid clutch designs fall into one of the two categories illustrated in Fig. 4.7. Although the concentric-cylinder design is simpler, both mechanically and with regard to the creation of the electric or magnetic field, the multiplate disk design can offer more area and thus more torque capability in a given volume.

Coulter et al. (1993) review some of the published work on clutches, summarizing factors such as the effects of electrical and viscous heating of the fluid. Among the applications reviewed therein is the use an ER clutch in a fall-intervention system intended to partially support an elderly person in the event that he or she falls while using an aid such as a walker. Carlson et al. (1996) examine several emerging commercial applications, and go into some detail about MR clutch and brake units designed for use in exercise equipment such as stationary bicycles, or for precision control of tension in manufacturing operations.

A high-speed traversing mechanism in which a belt carrying a load is coupled to one of two constant-speed, counter-rotating motors is described in Johnson et al. (1998), which includes the results of extensive numerical simulation work. The load is to be moved over a stroke of up to 250 mm at a steady-state belt velocity of 5 m/s, with a turn-around time of 10 ms. The simulations are used to compare modeling assumptions and to identify physical parameters, such as rotary inertias, that limit performance as

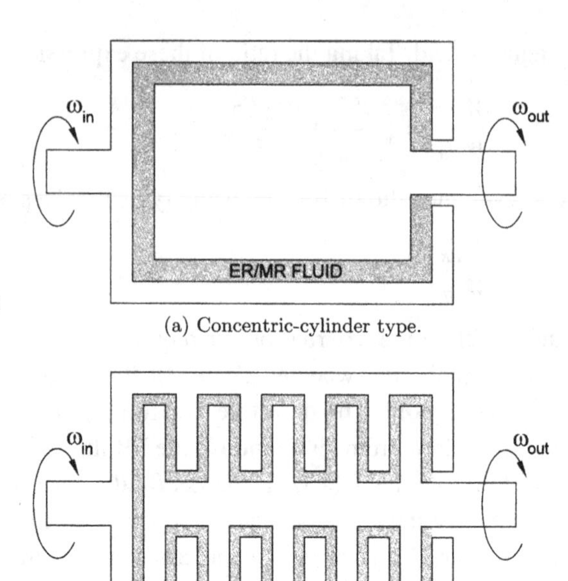

Figure 4.7. Two common ER or MR fluid clutch configurations.

measured by cycle time. Some experimental data are reported, along with practical experiences of electrical breakdown in aluminum and stainless-steel clutches.

The possibility of using a controllable-fluid clutch to transfer power from an automobile's engine to its drivetrain was raised first by Rabinow (1948) and has been examined repeatedly in the years since. At present, the consensus is that available ER fluids cannot support shear stresses of the magnitudes required in an automotive clutch, roughly 14 kPa (Carlson and Weiss, 1994). However, MR fluids display τ_y values several times this threshold, so the potential for an MR clutch suitable for use in a motor vehicle certainly exists. Other complications that might arise in this application include unacceptable levels of zero-field viscous drag (resulting in the transmission of some torque even when the clutch is nominally disengaged), and the requirement of dissipating the heat generated by shearing of the clutch fluid. Interestingly, these issues were raised by Rabinow in 1948.

4.3.2 Dampers

Another role in which ER and MR fluid devices are demonstrating great potential is that of dampers of mechanical vibration. The variable damper is a good example of a semi-active device: the force generated depends on the relative motion of the damper piston and cylinder, and very large forces can be developed with the input of only enough control energy to produce the desired fluid yield stress τ_y. Although aerospace applications such as landing gear oleos (Berg and Wellstead, 1998) and helicopter blade lag-mode dampers (Kamath, Werely, and Jolly, 1998; Marathe, Gandhi, and Wang, 1998) have been investigated, the greatest interest in ER and MR dampers appears to lie with the civil engineering community, which is actively studying them for vibration mitigation in buildings, bridges and other similarly massive structures. An important advantage to using semi-active, electrically controlled dampers of the type discussed here in civil structures that may be subject to wind loads or earthquake motions is that these dampers use so little electrical power that it is feasible to operate them from batteries if the main electricity supply fails. Furthermore, they are "fail safe" in that if the power to operate them is disrupted entirely, they perform as simple, passive dampers.

Dampers may be constructed in either a fixed-plate or sliding-plate configuration. An external, fixed-plate valve can be used to restrict the flow or fluid displaced by the damper piston, at the expense of some additional plumbing and inactive fluid volume. In dampers where the piston and cylinder form the walls of the gap, the piston plays the role of the sliding plate. The former arrangement has the advantage that the volume of the damper is constant, while in the latter an accumulator is usually incorporated to allow for the displacement of fluid by the piston rod. The accumulator introduces some stiffness, which may have to be modeled in design calculations and performance simulations.

In Carlson et al. (1996), a truck seat damper is described that is a direct replacement for a passive component. This MR damper has a 50 mm stroke and can generate forces up to 2 000 N. It can dissipate up to 600 W of mechanical power while requiring only 4 W at 1 A for control. This damper is used with a simple position feedback control.

The state of the art in controllable-fluid dampers as applied to civil structures is represented by the MR damper studied by Spencer et al. (1997). The device characterized in this paper has a 27 mm piston diameter with a gap of 0.5 mm between the piston and the cylinder wall. The gap is 15 mm in length, with 7 mm of this exposed to the magnetic field, corresponding to an active fluid volume of 0.3 ml. The field is applied radially, perpendicular to the flow. The damper contains 50 ml of MR fluid, works over a stroke of ± 2.5 cm, and can generate forces of up to 3 000 N. The coil has a resistance of 4 Ω and an inductance of 40 mH, resulting in an L/R time constant of 10 ms, and produces a field of 2 000 kA/m at a current of 1 A. The peak electrical power needed to control the damper is less than 10 W, so its operation from battery power is feasible.

The response of this damper to an imposed sinusoidal motion of frequency 2.5 Hz and amplitude 1.5 cm is shown in Fig. 4.8, where the curves corresponding to different magnet coil voltages may be seen to differ in amplitude but otherwise are quite similar. Note the well-defined but unusually shaped hysteresis loop around the origin of the force-velocity plot. Capturing this behavior can be important in predicting the dynamic response and energy dissipation of the damper or of a system depending on it.

The referenced paper reports the results of fitting various mechanical models to these test data, two examples of which we summarize here. The simplest model that reflects

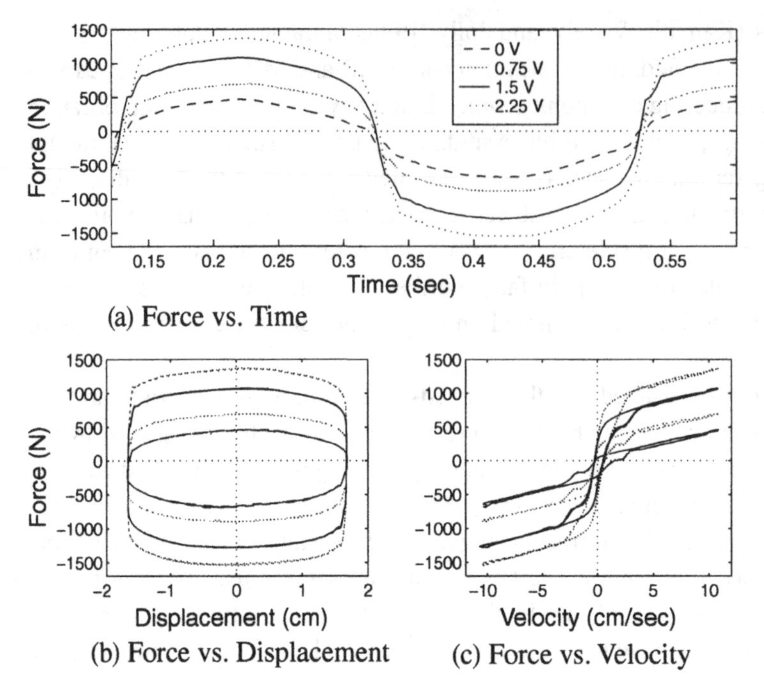

Figure 4.8. Experimentally measured force developed by the MR damper. (Reprinted, by permission, from Phenomenological model for magnet or heological dampers, *Journal of Engineering Mechanics*. Spencer et al., 1997. Copyright American Society of Civil Engineers.)

both the yield stress τ_y and the viscous stress $\eta\dot{\gamma}$ terms of the Bingham plastic model comprises a dashpot with coefficient c_0 in parallel with a Coulomb friction element with limit force f_c. For nonzero piston velocity $\dot{x}$, this model predicts a damper force of

$$f = f_c \operatorname{sgn} \dot{x} + c_0 \dot{x} + f_0, \tag{4.36}$$

where f_0 is included to account for the effect of the accumulator. Fitting the parameters of this model to the experimental data resulted in $f_c = 670\,\mathrm{N}$, $c_0 = 50\,\mathrm{N\,s/cm}$, and $f_0 = -95\,\mathrm{N}$. A comparison of the response predicted by using this model to the experimental data is shown in Fig. 4.9. Although the gross response is reasonably well approximated by this simple model, clearly visible discrepancies exist, especially at low velocities. Because the model predicts a one-to-one force-velocity relationship, it cannot reflect the experimentally observed hysteretic behavior at all.

This mechanical model was then systematically extended, building up to the one shown in Fig. 4.10. The Bouc-Wen block represents a widely used model of hysteretic behavior, and is described in terms of an evolutionary variable z governed by a first-order, nonlinear differential equation with three parameters that can be adjusted to alter the model's response. In the figure, c_0 represents the viscous damping dominant at larger velocities, while c_1 allows the model to follow the roll-off observed experimentally at low velocities; k_0 controls the model stiffness at large velocities, and the combination of k_1 and the initial displacement x_0 are associated with the nominal force of the accumulator. The reader is referred to the paper cited for details of fitting the fourteen

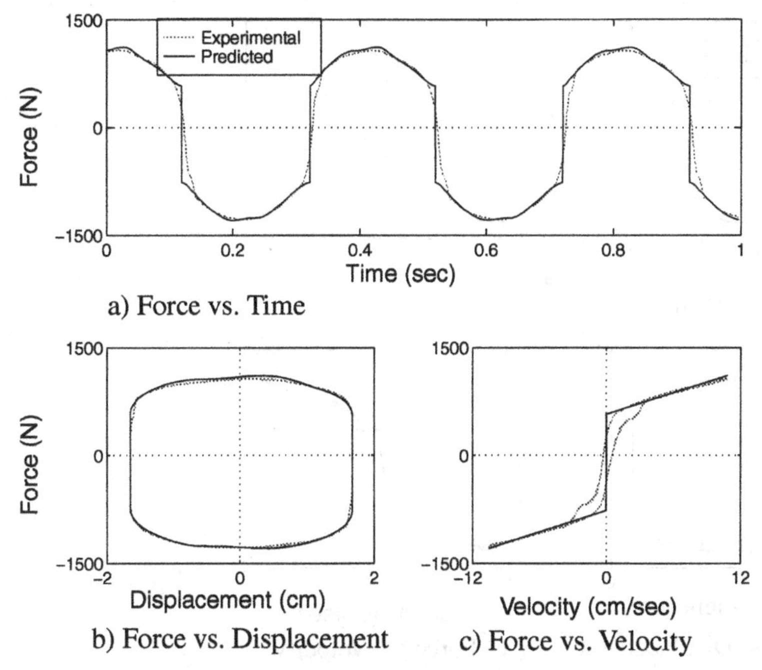

Figure 4.9. Comparison between experimental data and the response predicted by the Bingham model. (Reprinted, by permission, from Phenomenological model for magnet or heological dampers, *Journal of Engineering Mechanics*. Spencer et al., 1997. Copyright American Society of Civil Engineers.)

parameters of this model to the experimental data, but it is clear from Fig. 4.11 that the actual response, including the hysteresis loop, is much better represented by this model.

4.3.3 Other Applications

When ER and MR fluids are considered for use in smart structures, it is often in the form of a discrete device, such as the clutches and dampers discussed above. However, there is also interest in the use of controllable fluids as integral elements of structural systems. It is characteristic of many of these applications that they expose the energized fluid to strains below its yield strain (Coulter et al., 1993), and so the viscoelastic response of

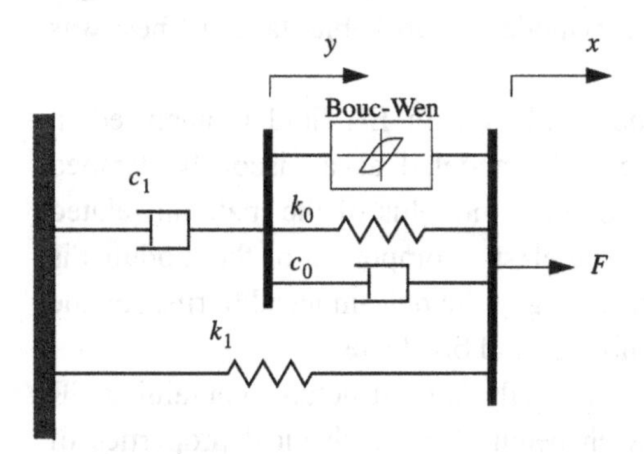

Figure 4.10. Elaborate mechanical model of the MR damper. (Reprinted, by permission, from Phenomenological model for magnet or heological dampers, *Journal of Engineering Mechanics*. Spencer et al., 1997. Copyright American Society of Civil Engineers.)

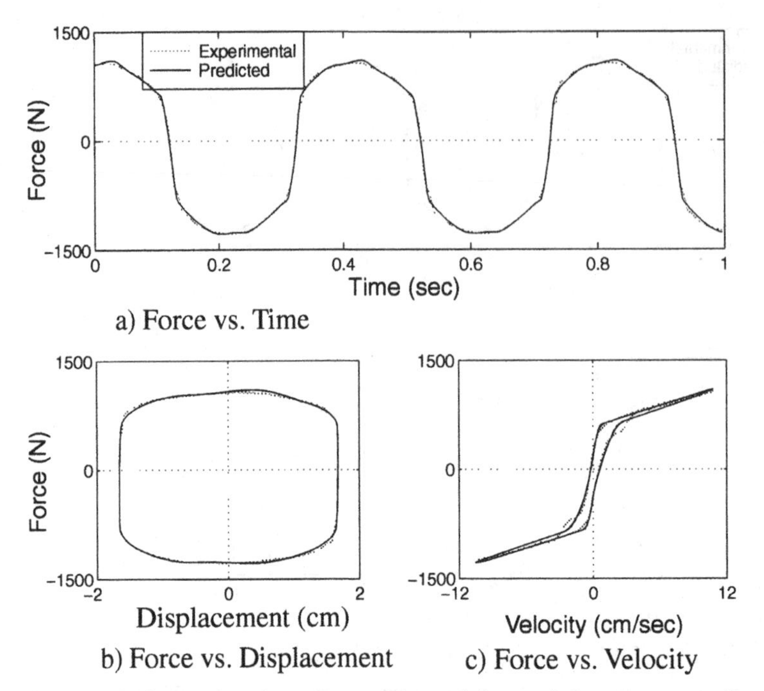

Figure 4.11. Comparison between experimental data and the response predicted by the model including a Bouc-Wen hysteresis block. (Reprinted, by permission, from Phenomenological model for magnet or heological dampers, *Journal of Engineering Mechanics*. Spencer et al., 1997. Copyright American Society of Civil Engineers.)

the unyielded controllable fluid is of greater interest here than in some other ER and MR devices.

Coulter et al. (1993) note that most designs in which an ER (or MR) fluid is distributed throughout a structural element can be classified as either shear or extensional configurations, depending upon the mode of deformation of the fluid. They go on to review analyses of both types. In the case of a shear-configuration beam containing ER fluid, the structural natural frequencies were found to vary linearly with the applied field, and the loss factors also increased with E. When the beam was analyzed by modeling the ER component like a damping treatment, qualitative agreement with experimental data was found, but quantitative results were "not definite." In an extensional configuration studied via a modified Timoshenko beam model, controllable static stiffness was demonstrated.

The free vibration of a cantilevered beam filled with ER fluid is analyzed in Choi, Sprecher, and Conrad (1990). The beam is modeled as a viscously damped single-degree-of-freedom system, and the complex modulus of the material related the the log decrement of the free response. The elastic component of the modulus is interpreted as resulting from interparticle stretching of the field-induced fibrils, and the loss component in terms of particle chain slippage and breakage.

In Lee and Cheng (1998), the finite-element analysis of structures containing ER fluid is reviewed, then applied to a sandwich beam. The mechanical properties of a rubber dam used to confine the ER fluid layer in the midplane of the specimen are incorporated using a rule of mixtures. A frequency-domain approach in which a swept-sine test is simulated, requiring an iterative solution at each frequency step, is used to calculate the frequency response of the beam. Good agreement is found with experimental data.

4.4 Summary

The potential of ER and MR fluids as the key element of an interface between electrical and mechanical subsystems is beginning to be realized in a variety of applications. The ability of these fluids to support a shear stress, either statically or dynamically, is at the heart of many practicable designs. After a long period of sporadic interest and activity, both ER and MR fluids are receiving steady attention from the research community, and commercial products are beginning to appear. Although the choice of a fluid must be made on a case by case basis, it appears that in some applications, especially those requiring high fluid yield stresses, MR fluids may be inherently better suited to meet the demands of performance and compatibility.

BIBLIOGRAPHY

Berg, C. D., and P. E. Wellstead. 1998. The non-linear effects of electrorheological fluids and their application in smart avionic systems. In *Proceedings of the SPIE Conference on Smart Structures and Integrated Systems* 3329:378–389, San Diego: SPIE.

Brooks, D. 1993. Applicability of simplified expressions for design with electrorheological fluids. *Journal of Intelligent Material Systems and Structures* 4:409–414.

Burton, S. A., Nicos Makris, I. Konstantopoulos, and P. J. Antsaklis. 1996. Modeling the response of ER damper: Phenomenology and emulation. *ASCE Journal of Engineering Mechanics* 122(9):897–906.

Carlson, J. D., D. M. Catanzarite, and K. A. St. Clair. 1996. Commercial magneto-rheological fluid devices. *International Journal of Modern Physics B* 10(23 & 24):2857–2865.

Carlson J. D., and Keith D. Weiss. 1994. A growing attraction to magnetic fluids. *Machine Design* 66(15):61–64.

Choi, Y., A. F. Sprecher, and H. Conrad. 1990. Vibration characteristics of a composite beam containing an electrorheological fluid. *Journal of Intelligent Material Systems and Structures* 1:91–104.

Coulter, J. P., Keith D. Weiss, and J. David Carlson. 1993. Engineering applications of electrorheological materials. *Journal of Intelligent Material Systems and Structures* 4:248–259.

Duclos, T. G. 1988. Design of devices using electrorheological fluids. *SAE Transactions, Journal of Materials* (Section 2) 97:2.532–2.536.

Dyke, S. J., B. F. Spencer, M. K. Sain, and J. D. Carlson. 1998. An experimental study of MR dampers for seismic protection. *Smart Materials and Structures* 7:693-703.

Gavin, H. P., R. D. Hanson, and F. E. Filisko. 1996a. Electrorheological dampers, Part I: Analysis and design. *ASME Journal of Applied Mechanics* 63:669–675.

Gavin, H. P., R. D. Hanson, and F. E. Filisko. 1996b. Electrorheological dampers, Part II: Testing and modeling. *ASME Journal of Applied Mechanics* 63:676–682.

Haiquing, G., Lim Mong Kim, and Tan Bee Cher. 1993. Influence of a locally applied electrorheological fluid layer on vibration of a simple cantilever beam. *Journal of Intelligent Material Systems and Structures* 4:379–384.

Johnson, A. R., W. A. Bullough, R. Tozer, and J. Makin. 1998. Simulation, performance and experimental validation of a high speed traversing/positioning mechanism using electrorheological clutches. In *Proceedings of the SPIE Conference on Smart Structures and Integrated Systems* 3329:402–413, San Diego: SPIE.

Kamath, G. M., N. M. Wereley, and M. R. Jolly. 1998. Characterization of semi-active magnetorheological helicopter lag mode dampers. In *Proceedings of the SPIE Conference on Smart Structures and Integrated Systems* 3329:356–377, San Diego: SPIE.

Kordonsky, W. 1993. Elements and devices based on magnetorheological effect. *Journal of Intelligent Material Systems and Structures* 4:65–69.

Lee, C.-Y., and C.-C. Cheng. 1998. Dynamic characteristics of a sandwich beam with embedded electro-rheological fluid. *Journal of Intelligent Material Systems and Structures* 9:60–68.

Makris, N., S. A. Burton, D. Hill, and M. Jordan. 1996. Analysis and design of ER damper for seismic protection of structures. *ASCE Journal of Engineering Mechanics* 122(10):1003–1011.

Marathe, S., F. Gandhi, and K. W. Wang. 1998. Helicopter blade response and aeromechanical stability with a magnetorheological fluid based lag damper. In *Proceedings of the SPIE Conference on Smart Structures and Integrated Systems* 3329:390–401, San Diego: SPIE.

Rabinow, J. 1948. The magnetic fluid clutch. *AIEE Transactions* 67:1308–1315, 1948.

Rao, N. N. 1977. *Elements of Engineering Electromagnetics*. Englewood Cliffs, NJ: Prentice-Hall.

Spencer, B. F., Jr., S. J. Dyke, M. K. Sain, and J. W. Carlson. 1997. Phenomenological model for magnetorheological dampers. *ASCE Journal of Engineering Mechanics* 123(3):230–238.

Spencer, B. F., Jr. and Michael K. Sain. 1997. Controlling buildings: A new frontier in feedback. *IEEE Control Systems* 17(6):19–35.

Spencer, B. F., Jr., G. Yang, J. D. Carlson, and M. K. Sain. 1998. "Smart" dampers for seismic protection of structures: A full-scale study. In *Proceedings of the Second World Conference on Structural control*, Kyoto, Japan.

Vieira, S. L., and A. C. F. de Arruda. 1998. Electrorheological fluids response under mechanical testing. *Journal of Intelligent Material Systems and Structure* 9:44–52.

Webb, N. 1990. Electrorheological fluids. *Chemistry in Britain* 26(4):338–340.

Weiss, K. D., J. David Carlson, and John P. Coulter. 1993a. Material aspects of electrorheological systems. *Journal of Intelligent Material Systems and Structures* 4:13–34.

Weiss, K. D., J. D. Carlson, and D. A. Nixon. 1994. Viscoelastic properties of magneto- and electro-rheological fluids. *Journal of Intelligent Material Systems and Structures* 5:772–775.

Weiss, K. D., T. G. Duclos, J. D. Carlson, M. J. Chrzan, and A. J. Margida. 1993. High strength magneto- and electro-rheological fluids. *SAE Transactions, Journal of Commercial Vehicles* (Section 2) 102:425–430.

Winslow, W. M. 1949. Induced fibration of suspensions. *Journal of Applied Physics* 20:1137–1140.

PROBLEMS

1. Calculate and plot the shear stress and velocity distributions through the thickness of the fluid layer in a fixed-plate controllable fluid device. Recall that during steady-state flow the fibrils break and reform almost entirely within layers near each wall. Consider separately the condition of

(a) $\tau < \tau_y$ (pre-yield) and

(b) $\tau = \tau_y$ (post-yield).

Compare the latter result to that for the flow of a conventional, viscous fluid in a similar passage.

2. Repeat Problem 1 for a shear-mode (moving-plate) device.

3. Fixed-plate ER valves have been built that produce pressure drops as high as 6.9 MPa (1 000 psi) (Coulter et al., 1993).

(a) Assuming the ER fluid used has a maximum yield stress of 3 kPa, find the minimum valve plate area needed to achieve this pressure drop in a 5-mm-diameter pipe. If the fluid gap is 0.5 mm, how large would such a device need to be (plate length and width) to offer the flow the same cross-sectional area as the pipe?

(b) Repeat the calculations of part (a) for an MR fluid of 80 kPa maximum yield stress.

4. Shown in Fig. 4.12 is an idealization of a controllable fluid mount discussed by Duclos (1988). The stiffness of this mount is dominated by the fixed top compliance (essentially a rubber "spring"), but the damping may be adjusted by controlling the electric or magnetic field across the orifice in the plate separating

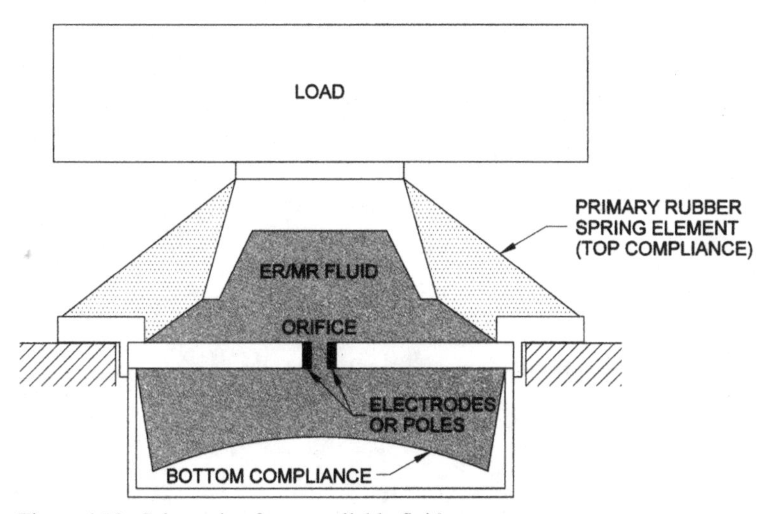

Figure 4.12. Schematic of a controllable fluid motor mount.

the top and bottom fluid chambers. Let the load have a mass of 120 kg and assume it is disturbed by a harmonic force of frequency ω.

(a) Sketch a lumped-parameter mechanical model of this system, including representative stiffness, mass, damping, and friction elements.

(b) Formulate the differential equation(s) governing the displacement response of this model.

(c) Making such assumptions as are necessary to produce a linear problem, calculate and plot the frequency response function relating displacement of the supported mass and the disturbance force in this structure.

(d) Numerically simulate the equations derived in (b) and quantify the effects of the simplifying assumptions introduced in (c). How do the mount performance and the error of the simple model vary with field strength?

5

Vibration Absorbers

5.1 Introduction

A frequently encountered problem in industry pertains to structures and structural components experiencing vibrations whose amplitude may escalate to dangerous levels leading to fatigue failure of the structure. Thus, buildings, bridges, airplanes, jet engine blades, and a wide variety of structures and their components are susceptible to vibration as they encounter time-dependent forces. The resulting time-dependent response of structures may vary over a range of frequencies present in the forcing function. The service life of these structures vibrating at a frequency will depend primarily upon the number of cycles accumulated at a given amplitude.

Elastic bodies responding to time-dependent external forces experience resonance at natural frequencies of the system. The basic parameters that govern the dynamic behavior of elastic systems are mass distribution, system compliance, and any damping present in the system. Mass distribution and system compliance lead to a single governing parameter (i.e., system frequencies). Thus, if a major vibration problem were to exist at a forcing frequency, a design choice can be to change the mass or compliance to place the vulnerable natural frequencies away from the forcing frequency. An alternate choice is to attempt to increase the levels of damping. Where neither of these choices is acceptable, external devices will be needed to overcome the vibratory fatigue problem.

Vibration reduction devices have been a subject of engineering research for a century now, and one of the most simple and practical devices, due to Frahm (Den Hartog, 1985), is shown in Fig. 5.1. In its simplest form, the Frahm absorber consists of an auxiliary undamped spring-mass system attached to the vibrating mass at a point where it is required to react the effective excitation force. With reference to Fig. 5.1, the equations of motion for the two degree-of-freedom system may be shown to be

$$m_1\ddot{x}_1 + (k_1 + k_2)x_1 - k_2x_2 = f_0e^{i\omega t}, \tag{5.1a}$$

$$m_2\ddot{x}_2 + k_2(x_2 - x_1) = 0. \tag{5.1b}$$

Without the attached absorber m_2, the vibratory amplitudes of the main mass m_1 may

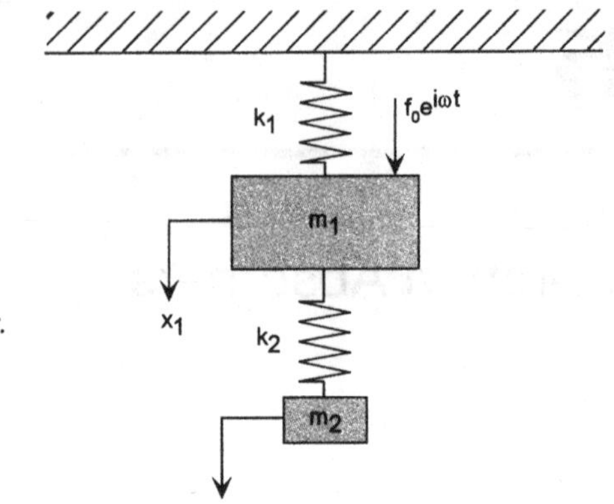

Figure 5.1. Frahm absorber.

reach larger and larger values as the forcing frequency ω approaches the natural frequency ω_1 ($\sqrt{k_1/m_1}$) of the main mass. Theoretically at $\omega = \omega_1$, the amplitude of m_1 is infinite and represents a resonant condition.

With the absorber attached, the solution to the differential equations (5.1) may be written as

$$x_1 = X_1 e^{i\omega t}, \tag{5.2a}$$

$$x_2 = X_2 e^{i\omega t} \tag{5.2b}$$

for steady state vibration. Substituting (5.2) into (5.1), we may show the response of the two masses to be

$$\frac{X_1}{X_{1_{ST}}} = \frac{1 - \frac{\omega^2}{\omega_2^2}}{\left(1 - \frac{\omega^2}{\omega_2^2}\right)\left(1 + \frac{k_2}{k_1} - \frac{\omega^2}{\omega_1^2}\right) - \frac{k_2}{k_1}} \tag{5.3a}$$

and

$$\frac{X_2}{X_{1_{ST}}} = \frac{1}{\left(1 - \frac{\omega^2}{\omega_2^2}\right)\left(1 + \frac{k_2}{k_1} - \frac{\omega^2}{\omega_1^2}\right) - \frac{k_2}{k_1}}, \tag{5.3b}$$

where $X_{1_{ST}}$ is the static displacement f_0/k_1 of the main mass. From (5.3a), it is clear that the vibratory amplitude of the main mass reaches zero if the natural frequency ω_2 of the absorber mass is tuned to be equal to the frequency of the forcing function at $\omega = \omega_1$ (i.e., $\omega_2 = \omega = \omega_1$). Thus, the vibratory motion of the main structure is "absorbed" at its natural frequency, resulting in a null. However, the absorber mass itself vibrates with the amplitude given by eq. (5.3b). The response characteristics of the main mass are displayed in Fig. 5.2 and clearly show two peaks corresponding to the two degrees-of-freedom and a null.

In summary, therefore, we may note that, when the natural frequency of the absorber mass is chosen to be equal to the frequency of the excitation force, then the main mass does not vibrate at all at this frequency and is said to attain a null. Although Frahm absorbers are quite popular in use because of their simplicity, their effectiveness is

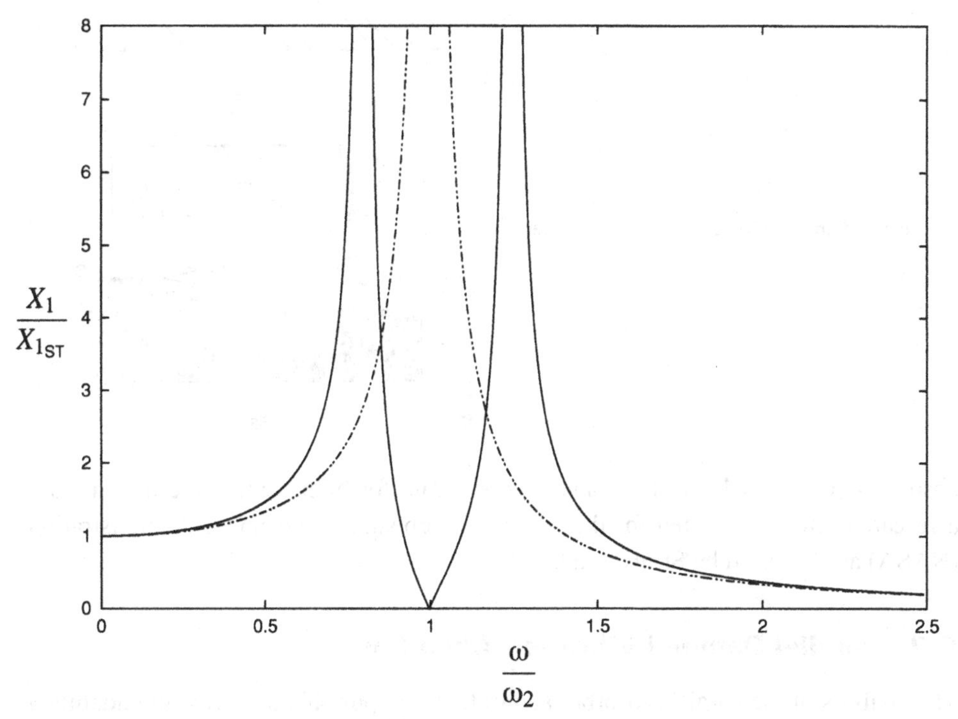

Figure 5.2. Response of the undamped main mass with no absorber (broken line) and with an undamped conventional vibration absorber (solid line).

rather limited to situations where the excitation frequency is nearly constant. In fact, the addition of an absorber mass introduces another degree of freedom to the system and thus another resonant condition which might do more harm than good.

However, the main drawback of the Frahm absorber lies in the narrow bandwidth of excitation frequencies within which the absorber is effective. In general, the excitation frequency varies over a range that renders the conventional absorbers essentially useless. The purpose of many investigations that have followed since the introduction of Frahm absorbers has been either:

1. To improve the effectiveness of the conventional absorber by suitable modification, or
2. To invent entirely different and better devices in the hope of replacing the conventional absorber.

Gyroscopic vibration absorbers and impact dampers are but a few of the new devices that belong to the latter group. A common modification considered so far in the former group is the addition of damping to the absorber mass. A further modification of the conventional absorber was considered by Srinivasan (1968) and represents an open-loop type of smartness. Such a modification consists of adding, in parallel, a subsidiary undamped absorber mass in addition to the damped absorber mass (Fig. 5.3). This modification, designated as parallel damped vibration absorber, is discussed in the section below, followed by a section devoted to the analysis of gyroscopic vibration

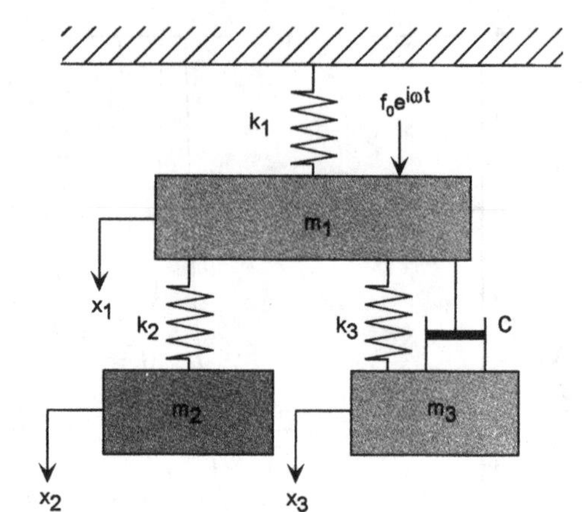

Figure 5.3. Parallel damped vibration absorber.

absorbers (GVAs). Much of the analyses and data for both devices are drawn from a research effort conducted for the National Aeronautics and Space Administration (NASA) and reported by Srinivasan (1968).

5.2 Parallel Damped Vibration Absorber

The analysis of the parallel absorber shows that it is possible to obtain an undamped antiresonance in a dynamic absorber system, which exhibits a well-damped resonance. Although the bandwidth of frequencies between the damped peaks is not significantly increased, the amplitudes of the main mass are considerably smaller within the operational range of the absorber. The damped absorber mass and the main mass attain null simultaneously so that the vibratory force is transmitted entirely to the undamped absorber. A comparison of the results with those of the conventional absorber indicates that the parallel damped dynamic vibration absorber has definite advantages over the conventional damped vibration absorber.

5.2.1 Analysis

The analysis that follows consists mainly of:

1. The derivation of the governing equations of motion, and
2. Derivation of the condition for the amplitude of the main mass to be independent of the damping ratio

The latter condition provides the frequencies at which the amplitudes of the main mass are independent of the damping ratio c/c_c. In addition, for the particular case of practical interest (i.e., when the absorber masses and the springs have the same value), the so-called favorable tuning has been determined. Favorable tuning refers to the frequency at which the absolute value of the amplitude is independent of damping. Under this favorable tuning, the mass ratio required to provide the greatest spread between the frequencies is determined. Also, the equation that provides the optimum damping ratio

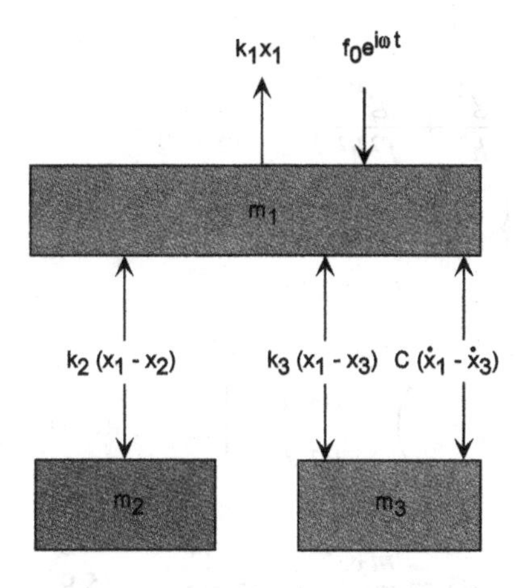

Figure 5.4. Forces acting on each mass.

has been derived and represents a condition at which the slope of the response curve is zero.

The forces acting on each mass of the system under consideration are shown in Fig. 5.4. The main mass is assumed to be subjected to the action of a periodic force $f_0e^{i\omega t}$. Only steady-state response is considered.

From Fig. 5.4, the equations of motion can be readily written as

$$m_1\ddot{x}_1 + k_1x_1 + k_2(x_1 - x_2) + k_3(x_1 - x_3) + C(\dot{x}_1 - \dot{x}_3) = f_0e^{i\omega t}, \tag{5.4a}$$

$$m_2\ddot{x}_2 + k_2(x_2 - x_1) = 0, \tag{5.4b}$$

$$m_3\ddot{x}_3 + k_3(x_3 - x_1) + C(\dot{x}_3 - \dot{x}_1) = 0. \tag{5.4c}$$

For the steady-state response, the solutions x_1, x_2, and x_3 may be assumed as

$$x_1 = X_1e^{i\omega t}, \tag{5.5a}$$

$$x_2 = X_2e^{i\omega t}, \tag{5.5b}$$

$$x_3 = X_3e^{i\omega t}. \tag{5.5c}$$

Substituting (5.5) in (5.4), a set of simultaneous equations in the unknowns X_1, X_2 and X_3 may be obtained. These equations may be represented in the form of a symmetric matrix as

$$\begin{bmatrix} k_1 + k_2 + k_3 - m_1\omega^2 + Ci\omega & -k_2 & -k_3 + Ci\omega \\ -k_3 & k_2 - m_2\omega^2 & 0 \\ -k_3 + Ci\omega & 0 & k_3 - m_3\omega^2 + Ci\omega \end{bmatrix} \times \begin{Bmatrix} X_1 \\ X_2 \\ X_3 \end{Bmatrix} = \begin{Bmatrix} f_0 \\ 0 \\ 0 \end{Bmatrix}. \tag{5.6}$$

From eq. (5.6), $X_1/X_{1_{\text{ST}}}$ may be derived as shown below:

$$\frac{X_1}{X_{1_{\text{ST}}}} = \frac{1}{D}\left(1 - \frac{g^2}{f^2} - \frac{g^2}{h^2} + \frac{g^4}{f^2h^2}\right) + i2C_r\frac{g}{f^2}\left(1 - \frac{g^2}{h^2}\right), \tag{5.7}$$

where

$$D = \left[\left\{(1+\mu_2 h^2+\mu_3 f^2-g^2)\left(1-\frac{g^2}{f^2}-\frac{g^2}{h^2}+\frac{g^4}{f^2h^2}\right)\right.\right.$$
$$\left.+\mu_2 h^2\left(\frac{g^2}{h^2}-1\right)+\mu_3 f^2\left(\frac{g^2}{h^2}-1\right)\right\}$$
$$+i2C_r g\mu_3\left\{\left(1-\frac{g^2}{f^2}-\frac{g^2}{h^2}+\frac{g^4}{f^2h^2}\right)\right.$$
$$\left.\left.+\left(1-\frac{g^2}{h^2}\right)\left(\frac{\mu_2}{\mu_3}\frac{h^2}{f^2}+\frac{1}{\mu_3}\frac{1}{f^2}-\frac{1}{\mu_3}\frac{g^2}{f^2}-1\right)-\frac{\mu_2}{\mu_3}\frac{h^2}{f^2}\right\}\right] \tag{5.8}$$

and

$$g=\frac{\omega}{\omega_1},\quad f=\frac{\omega_3}{\omega_1},\quad h=\frac{\omega_2}{\omega_1},\quad \mu_2=\frac{m_2}{m_1},\quad \mu_3=\frac{m_3}{m_1},\quad C_r=\frac{C}{C_c}, \tag{5.9}$$

ω is the frequency of the forcing function, and ω_1, ω_2, and ω_3 are the natural frequencies of the main mass, undamped absorber mass, and damped absorber mass, respectively. Similarly, $X_2/X_{1_{ST}}$ and $X_3/X_{1_{ST}}$ may be represented in dimensionless form as

$$\frac{X_2}{X_{1_{ST}}}=\frac{1}{D}\left[\left(1-\frac{g^2}{f^2}\right)+i2C_r\frac{g}{f^2}\right] \tag{5.10}$$

and

$$\frac{X_3}{X_{1_{ST}}}=\frac{1}{D}\left[\left(1-\frac{g^2}{f^2}\right)\left(1+i2C_r\frac{g}{f^2}\right)\right]. \tag{5.11}$$

A condition that renders $X_1/X_{1_{ST}}$ to be independent of the damping ratio can be shown to be

$$g^6(2+\mu_3)-g^4\{\mu_3(h^2+2f^2)+2\mu_2h^2+2(f^2+h^2)+2\}$$
$$+2g^2\{f^2h^2(1+\mu_2+\mu_3)+(f^2+h^2)\}-2f^2h^2=0. \tag{5.12}$$

Equation (5.12) is a cubic in g^2, and for given values of μ_2, μ_3, f, and h it may be solved to obtain the values of g at which $X_1/X_{1_{ST}}$ is independent of C_r. Using these values of g, the amplitudes of the main mass may be computed from eq. (5.7).

5.2.2 A Special Case

Although further analysis may proceed with all the parameters in their most general form, it is found that considerable simplification in computation results if the absorber masses and the corresponding spring rates are assumed to be equal. Also, such an assumption leads to a case of practical interest. With $\mu_2=\mu_3=\mu$ and $k_2=k_3$, eq. (5.12) reduces to

$$g^6(2+\mu)-g^4\{f^2(4+5\mu)+2\}+2g^2\{2f^2+f^4(2+\mu)\}-2f^4=0. \tag{5.13}$$

With a little algebraic manipulation, eq. (5.13) may be shown to be

$$\{g^4(2+\mu)-2g^2(2\mu f^2+f^2+1)+2f^2\}(g^2-f^2)=0. \tag{5.14}$$

Because $g = f$ corresponds to the null, the dimensionless frequencies, g, at which the amplitudes $X_1/X_{1_{ST}}$ are independent of damping are given by

$$g^4 - \frac{2g^2(2\mu f^2 + f^2 + 1)}{2+\mu} + \frac{2f^2}{2+\mu} = 0. \tag{5.15}$$

Equation (5.15) is a quadratic in g^2 and provides the two required values of g (say g_1 and g_2). Using the values of g_1 and g_2, the corresponding values of the ratio $X_1/X_{1_{ST}}$ may be computed from a simplified equation obtained from eq. (5.7), i.e.,

$$\left(\frac{X_1}{X_{1_{ST}}}\right)_{g_1,g_2} = \frac{f^2 - g^2}{g^4 - g^2(1 + f^2 + 2\mu f^2) + f^2}. \tag{5.16}$$

Equation (5.16) is obtained from the general expression (5.7) by making the assumptions $\mu_2 = \mu_3 = \mu$, $f = h$, and by letting $C_r = 0$. The latter assumption is valid because at g_1 and g_2, $X_1/X_{1_{ST}}$ is independent of C_r.

The amplitudes at g_1 and g_2, as computed from eq. (5.16), are, in general, not equal. In order to avoid the necessity to refer to two different amplitudes and make comparisons at every stage, the absolute value of these two amplitudes may be forced to be equal at g_1 and g_2. Because it is not apparent whether the amplitudes at g_1 and g_2 are of the same sign, the general requirement may be written as

$$\frac{f^2 - g_1^2}{(f^2 - g_1^2)(1 - g_1^2 + 2\mu f^2) - 2\mu f^4} = \pm \frac{f^2 - g_2^2}{(f^2 - g_2^2)(1 - g_2^2 + 2\mu f^2) - 2\mu f^4}. \tag{5.17}$$

The solution of eq. (5.17) provides the value of the tuning (f) such that the absolute amplitude of the main mass at g_1 is the same as that at g_2. Omitting the details of calculation, the required equations may be shown to be

$$(f^2 - g_2^2)(f^2 - g_1^2) + 2\mu f^4 = 0 \tag{5.18}$$

with the + sign and

$$(f^2 - g_1^2)(f^2 - g_2^2)(4\mu f^2 - g_1^2 - g_2^2 + 2) - 2\mu f^4(2f^2 - g_1^2 - g_2^2) = 0 \tag{5.19}$$

with the − sign.

Clearly, only one of these equations is valid. The valid equation may be found as follows. The roots g_1 and g_2 of the quadratic eq. (5.15) satisfy the conditions

$$g_1^2 + g_2^2 = \frac{4\mu f^2 + 2f^2 + 2}{2+\mu} \tag{5.20a}$$

and

$$g_1^2 g_2^2 = \frac{2f^2}{2+\mu}. \tag{5.20b}$$

Equation (5.18) may be written as

$$(g_1^2 + g_2^2)f^2 - g_1^2 g_2^2 - (1 + 2\mu)f^4 = 0. \tag{5.21}$$

Substitution of (5.20a) and (5.20b) in (5.21) leads to

$$f^4(2\mu^2 + \mu) = 0. \tag{5.22}$$

Equation (5.22) is satisfied only when the absorbers have an uncoupled natural frequency of zero ($f = 0$), which is a trivial condition. Thus, the amplitudes (at g_1 and g_2) are opposite in sign, and the tuning f required to make them equal is obtained from eq. (5.19), written in the form

$$\begin{aligned}(g_1^2 + g_2^2)\{(g_1^2 + g_2^2)f^2 - 2\mu f^4 - g_1^2 g_2^2 - 2f^2 - f^4\}\\ + 2g_1^2 g_2^2(1 + 2\mu f^2) + 2f^4 = 0.\end{aligned} \tag{5.23}$$

As before, substituting the expressions for $g_1^2 + g_2^2$ and $g_1^2 g_2^2$ from eqs. (5.20) in eq. (5.23), the resulting equation may be shown to be

$$f^2 = \frac{1 - \mu}{(1 + 2\mu)^2}. \tag{5.24}$$

The required tuning, the so-called favorable tuning, which gives equal amplitudes can be calculated from eq. (5.24).

5.2.2.1 The Optimum Case

In discussing the problem of damped vibration absorbers, it is customary to determine "optimum amplitude" and the corresponding "optimum damping." The optimum damping is defined as the damping required to obtain a zero slope of the response curve at g_1 and g_2, and the resulting amplitude at g_1 and g_2 is termed the optimum amplitude. Den Hartog (1985), in his analysis of a damped absorber, comments that the calculations involved in computing the optimum damping are "long and tedious." For the systems described here, the calculations are much longer and more tedious. The optimum C_r is determined from the expression

$$\left\{\frac{d}{dg}\left(\frac{X_1}{X_{1_{ST}}}\right)\right\}_{g_{1,2}} = 0. \tag{5.25}$$

The calculations shown below are for the case when $\mu_2 = \mu_3 = \mu$ and $f = h$. The roots g_1 and g_2 are, as before, the values of g at which $X_1/X_{1_{ST}}$ is independent of C_r:

$$\left(\frac{X_1}{X_{1_{ST}}}\right)^2 = \frac{A_1^2 + B_1^2}{C_1^2 + D_1^2}, \tag{5.26}$$

where

$$A_1 = (f^2 - g^2)^2, \tag{5.27a}$$

$$B_1 = 2C_r g(f^2 - g^2), \tag{5.27b}$$

$$C_1 = (f^2 - g^2)\{(1 - g^2)(f^2 - g^2) - 2\mu f^2 g^2\}, \tag{5.27c}$$

$$D_1 = 2C_r g\{g^4(1 + \mu) - f^2 g^2(1 + 2\mu) + (f^2 - g^2)\}. \tag{5.27d}$$

Then

$$\frac{d}{dg}\left(\frac{X_1}{X_{1_{ST}}}\right)^2 = (C_1^2 + D_1^2)\left\{A_1\frac{dA_1}{dg} + B_1\frac{dB_1}{dg}\right\} - (A_1^2 + B_1^2)\left\{C_1\frac{dC_1}{dg} + D_1\frac{dD_1}{dg}\right\} \quad (5.28)$$

and therefore

$$\left[(A_1A_1' + B_1B_1') - \left(\frac{X_1}{X_{1_{ST}}}\right)^2 (C_1C_1' + D_1D_1')\right] = 0, \quad (5.29)$$

where the primes indicate derivatives with respect to g.

In the following, the terms appearing in (5.29) will be listed in order, in their final form, and all details of calculation will be omitted.

$$A_1A_1' = -4g(f^2 - g^2)^3, \quad (5.30)$$

$$B_1B_1' = 4C_r^2 g(f^2 - g^2)(f^2 - 3g^2), \quad (5.31)$$

$$\begin{aligned} C_1C_1' = {} & 6g^{11} - 5g^9(2 + 4f^2 + 4\mu f^2) + 4g^7(4\mu f^2 + 8f^2 + 12\mu f^4 + 6f^4 + 1) \\ & - 12g^5(3\mu f^4 + 3\mu f^6 + 2\mu^2 f^6 + 3f^4 + f^6 + f^2) \\ & + 2g^3(6f^4 + 8f^6 + 12\mu f^6 + f^8 + 4\mu f^8 + 4\mu^2 f^8) \\ & - 2g(2f^6 + f^8 + 2\mu f^8), \end{aligned} \quad (5.32)$$

$$\begin{aligned} D_1D_1' = {} & 4C_r^2 \{5g^9(1 + \mu)^2 - 8g^7(f^2 + 3\mu f^2 + 2\mu^2 f^2 + 1 + \mu) \\ & + 3g^5(f^4 + 4\mu f^4 + 4f^2 + f\mu^2 f^4 + 6\mu f^2 + 1) \\ & - 4g^3(f^4 + 2\mu f^4 + f^2) + f^4 g\}. \end{aligned} \quad (5.33)$$

Substituting these expressions in eq. (5.29), an equation for optimum damping ratio ($C_{r_{OPT}}$) can be obtained in the form

$$C_{r(OPT)} = \sqrt{\frac{y^2h_1 - h_2}{h_3 - y^2h_4}} \quad (5.34)$$

where h_1, h_2, h_3, and h_4 are known functions of f and g, and $y = X_1/X_{1_{ST}}$ computed at either g_1 or g_2. The value of $C_{r(OPT)}$ obtained from eq. (5.25) at g_1 is, in general, different from that at g_2. An average value of $C_{r(OPT)}$ is therefore proposed as the required optimum damping ratio.

5.3 Numerical Results

The responses of the main mass and the absorber masses have been represented graphically as functions of the frequency ratio g with damping ratio C_r as a parameter. Also, the phase angles of the main mass shown are computed and represented graphically. The response for $f = 1$ and $f = f_f$ are shown in Figs. 5.5 and 5.6, respectively. In

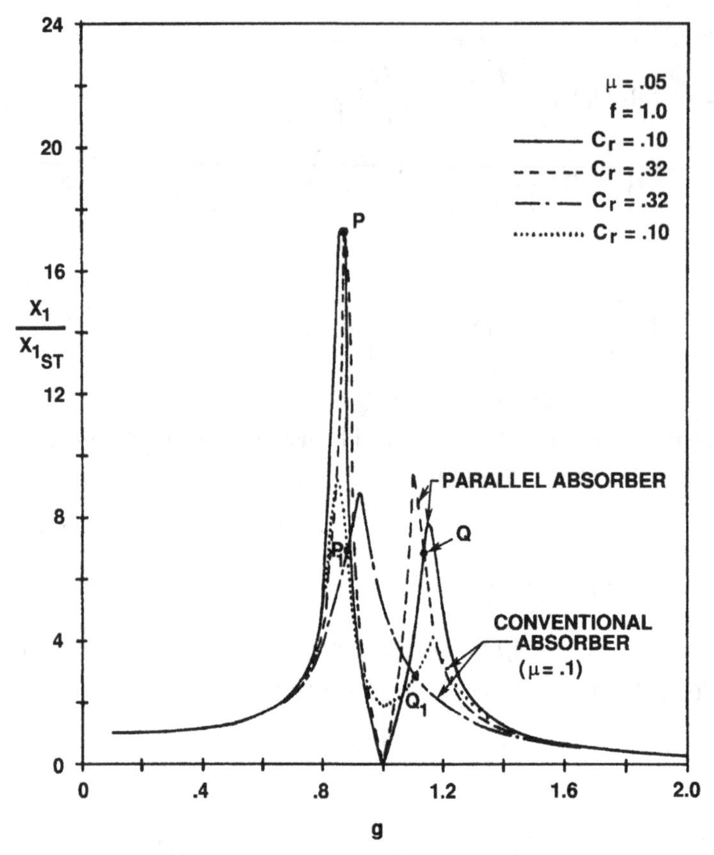

Figure 5.5. Amplitudes of main mass when the absorbers and main mass are tuned to the same frequency (Srinivasan, 1968).

order to judge the effectiveness of the parallel vibration absorber, the responses of the conventional absorber are compared with those of the corresponding parallel vibration absorber. The latter is defined as the parallel vibration absorber whose absorber masses are each equal to one-half of the absorber mass of the conventional absorber.

An examination of Fig. 5.5 shows the principal features of the comparison. The introduction of an undamped absorber mass, in addition to the damped absorber mass, has made it possible to obtain an undamped antiresonance in a dynamic absorber system that exhibits two damped resonances. This is an expected result and is decidedly an advantage. However, the amplitudes increase rather sharply for small changes in the frequency ratio g, thus retaining the disadvantages of the conventional damped absorbers. Therefore, both the absorbers permit only very small tolerances in the change of frequency ratio g. Nevertheless, the parallel absorber appears to be superior to the conventional damped absorber if a comparison is made between the response curves for a damping ratio such as $C_r = 0.32$. The conventional absorber has, for this ratio of C_r, prohibitively large amplitudes within the operational range of the vibration absorber. Also, the characteristic feature of the response curve of the conventional absorber changes significantly in that the two smaller peaks for a low damping ratio such as for

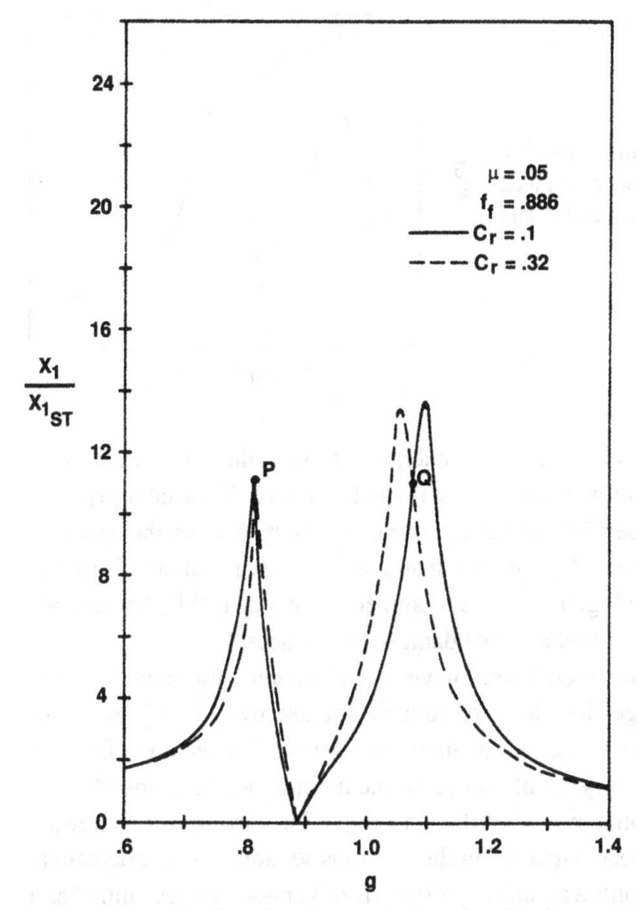

Figure 5.6. Amplitudes of main mass when the absorbers and main mass are favorably tuned (Srinivasan, 1968).

$C_r = 0.1$ tend to merge to a single, but larger, peak when C_r is increased. On the other hand, in the case of the parallel absorber, the characteristic features (i.e., two damped peaks and a null) remain intact when the damping is changed and the amplitudes within the operational range of frequencies are considerably smaller for higher damping ratios. Even for low damping ratios, it may be observed that the amplitudes in the narrow range between the peaks, P_1 and Q_1, are smaller.

In view of the fact that the above advantages are obtainable by merely assigning one-half of the absorber mass of the conventional absorber as the undamped mass of the parallel absorber, the device is obviously preferable.

5.4 Gyroscopic Vibration Absorbers

Both the conventional and the parallel absorbers discussed above are designed to transfer energy from the main structures (main mass) to subsidiary structures (absorber masses).

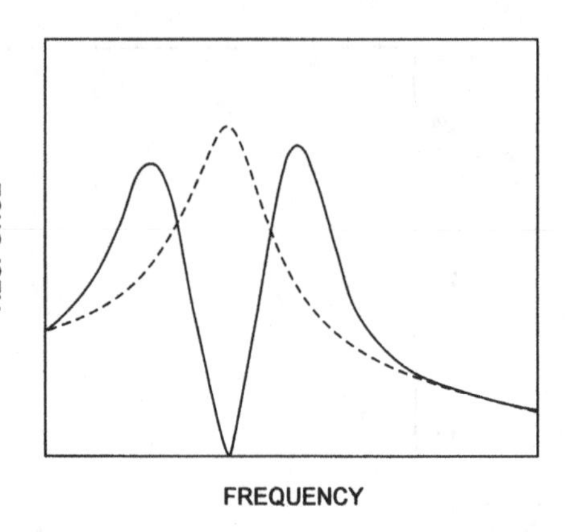

Figure 5.7. General response of a structure alone (dashed line) and with a conventional damped vibration absorber (solid line) (Flannelly and Wilson, 1967).

The latter vibrate in the process of absorbing energy but the main mass or structure itself stops vibrating or its amplitude of vibration is vastly reduced. Translatory motion of the main structure is, in a sense, transferred to translatory motion of the absorber masses. The mathematical relationships are linear and the devices are suitable for use at a single frequency, as shown in Fig. 5.7, where a damped resonance is fully "absorbed" by an absorber mass leading to a null and two damped resonances.

The devices discussed in this section are known as gyroscopic vibration absorbers and have the unique advantage that they are suitable for use over a wide range of forcing frequencies. Furthermore, they are distinguished by another feature: vibratory energy from translatory or other types of motion of the main structure is absorbed by the device resulting in vibrational motion of the gyroscope. The devices are somewhat more complex than the absorbers discussed in the previous section. The mathematical relationships are essentially nonlinear, although linearized versions are examined and may be relevant in certain applications.

The idea of utilizing gyroscopic dynamics resulting in a completely inertial/conservative means of reacting a sinusoidal force originated with an invention by W. G. Flannelly (Flannelly and Wilson, 1967) designated the *gyroscopic vibration absorber*. In an effort to develop devices to eliminate vibration in helicopters, Flannelly's ingenuity led to a variety of schemes he sketched out for further analysis, experimentation, and development. Students of dynamics would benefit greatly by a thorough study of the cited report. A few of the schematics are reproduced below and offer a rich source of inspiration for further research in the field of vibration absorption (see Figs. 5.8 through 5.10). The analysis presented in the section below is developed for the configuration in which the gyro wheel, the drive system, and the cross pivots all rotate in unison. Flannelly named it the perissogyro vibration absorber.

In these devices, a wide band of null frequencies (antiresonance) becomes possible through synchronization of the speed of the gyrowheel with the forcing frequency as shown schematically in Fig. 5.11. This feature is most desirable in applications where the frequency of the forcing function is not constant. A particular application in which the system characteristics vary continuously over time is in machining components of a wide variety of equipment. Machining is a dynamic process, and the continuous

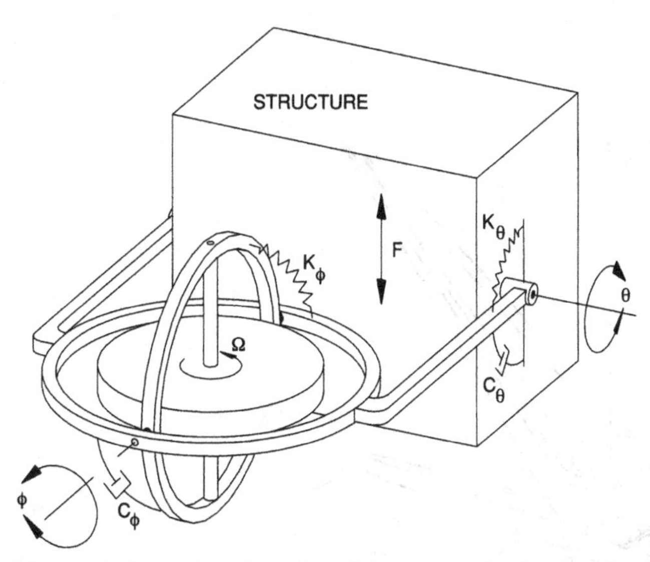

Figure 5.8. General configuration of the gyroscopic vibration absorber (after Flannelly and Wilson, 1967).

removal of material during the process changes the mass and compliance of the workpiece. The cutting tool may begin to vibrate and, depending upon the system characteristics, the machining process may exhibit an instability known as chatter. Either the speed of cutting or the width of cut needs to be adjusted downward to obtain smooth machining. This inevitably leads to increased machining time and costs. Gyroscopic vibration absorbers can be ideal devices in such applications and need to be developed further for use in several machining processes such as boring, drilling, and milling. In

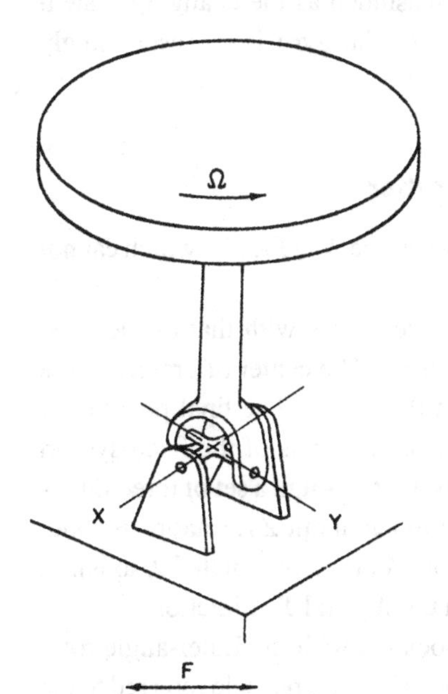

Figure 5.9. Perissogyro vibration absorber (Flannelly and Wilson, 1967).

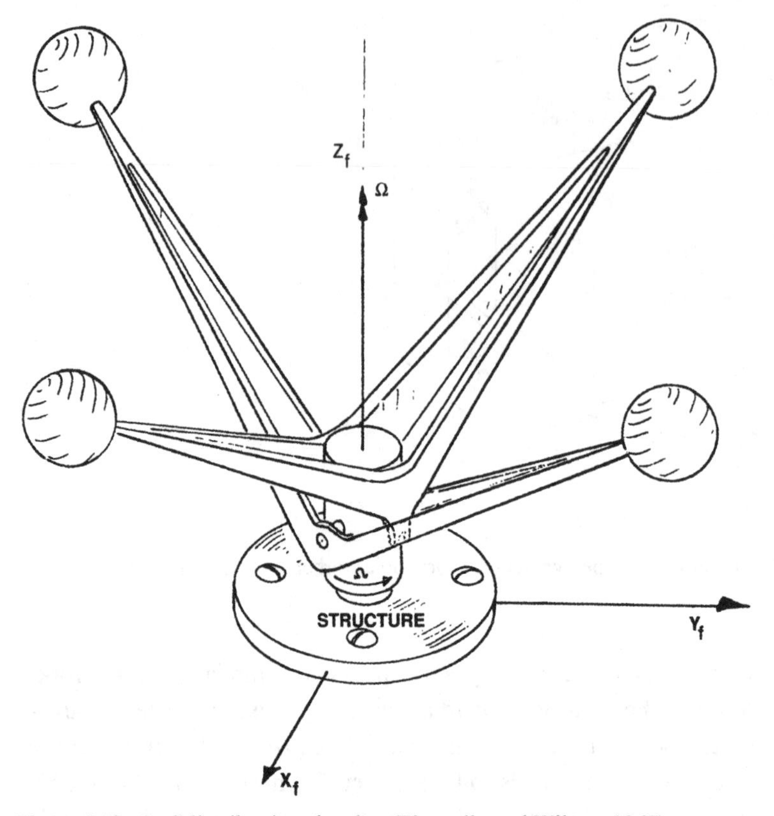

Figure 5.10. Coriolis vibration absorber (Flannelly and Wilson, 1967).

the following, we present details of one type of gyroscopic absorber. Such absorbers belong to the class of closed-loop smart structures inasmuch as the changing system frequency is continuously monitored and the speed of the gyro is correspondingly synchronized.

5.4.1 Analysis: Perissogyro Vibration Absorber

Figure 5.12 represents a fixed coordinate system of reference X_f, Y_f, Z_f, which cannot rotate but can translate in the X_f, Y_f directions.

The coordinate system X_b, Y_b, Z_b, whose origin coincides with that of the fixed system, is fixed to the body of the gyro and moves with it. The center of gravity of the rigid body is assumed to be located at h units from the origin. The final position of an element of mass dm at any instant may be expressed in terms of the fixed system coordinates. The body assumes its final position at instant t through a set of three Euler-angle rotations given in the following sequence: rotation ψ about Z_f, rotation θ about Y_1 (i.e., the rotated Y axis), and finally rotation φ about X_b (i.e., the rotated X axis). In addition, the rigid body may undergo translations in the X_f and Y_f directions.

Representing the matrices of transformation associated with the Euler-angle rotations ψ, θ, and φ by $[\Psi]$, $[\Theta]$, and $[\Phi]$, the coordinates of a mass particle referred to the

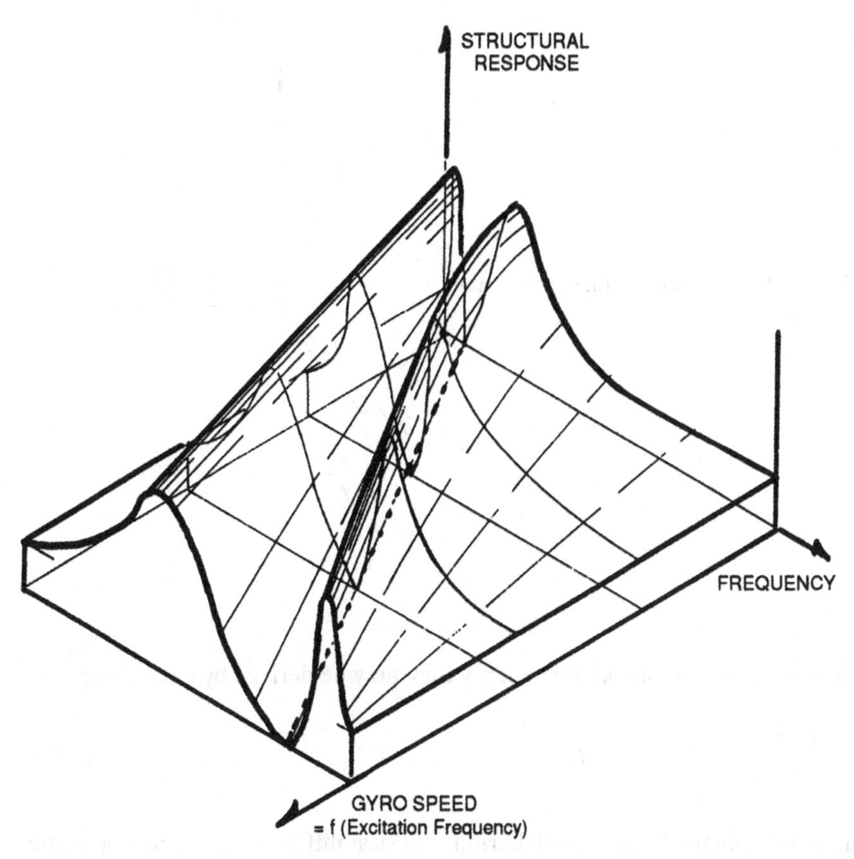

Figure 5.11. General response of a structure with synchronous vibration absorber (Flannelly and Wilson, 1967).

fixed system may be expressed in terms of the body coordinates by the matrix equation

$$\begin{Bmatrix} X_f \\ Y_f \\ Z_f \end{Bmatrix} = [[\Psi][\Theta][\Phi]] \begin{Bmatrix} X_b \\ Y_b \\ Z_b \end{Bmatrix}, \tag{5.35}$$

where

$$[\Psi] = \begin{bmatrix} \cos\psi & -\sin\psi & 0 \\ \sin\psi & \cos\psi & 0 \\ 0 & 0 & 1 \end{bmatrix}, \tag{5.36}$$

$$[\Theta] = \begin{bmatrix} \cos\theta & 0 & \sin\theta \\ 0 & 1 & 0 \\ -\sin\theta & 0 & \cos\theta \end{bmatrix}, \tag{5.37}$$

and

$$[\Phi] = \begin{bmatrix} 1 & 0 & 0 \\ 0 & \cos\varphi & -\sin\varphi \\ 0 & \sin\varphi & \cos\varphi \end{bmatrix}. \tag{5.38}$$

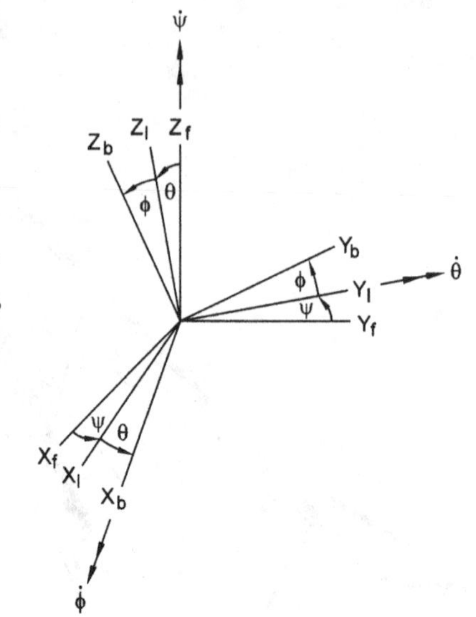

Figure 5.12. Coordinate transformation (Srinivasan, 1968).

The expression for the total kinetic energy may now be derived by calculating

$$T = \frac{1}{2}\int \left(\dot{x}_f^2 + \dot{y}_f^2 + \dot{z}_f^2\right) dm. \tag{5.39}$$

The calculations to obtain T are rather laborious but straightforward. The body axes are assumed to be along the principal axes of inertia of the rigid body. Thus, in calculating the kinetic energy integral from eq. (5.39), all the products of inertial terms are assumed to identically zero.

The potential energy stored in the system is assumed to be solely due to the spring rates in the pivots and along the X_f and Y_f directions; that is, the contribution to the total potential energy from gravitational forces is neglected in the analysis. Thus, the expression for V may be written as

$$V = \tfrac{1}{2}\{K_\theta\theta^2 + K_\varphi\varphi^2 + K_x X_0^2 + K_y Y_0^2\} \tag{5.40}$$

where K_θ, K_φ are the spring rates in the pivots along the θ, φ directions, and K_x, K_y represent the spring rates along the X, Y directions.

The governing equations of motion in terms of the generalized coordinates ψ, θ, φ, X_0, and Y_0 may now be derived from the Lagrangian L, using

$$\frac{d}{dt}\frac{\partial L}{\partial \dot{q}_i} - \frac{\partial L}{\partial q_i} = Q_i \tag{5.41}$$

where q_i is a generalized coordinate and Q_i the corresponding generalized force. The resulting equations are a set of coupled nonlinear ordinary differential equations and are given below.

$$\begin{aligned}&\ddot{\psi}\left(C_\theta^2 S_\varphi^2 I_y + S_\theta^2 I_x + C_\theta^2 C_\varphi^2 I_z\right) + \ddot{\theta} C_\theta C_\varphi S_\varphi (I_y - I_z) - \ddot{\varphi} S_\theta I_x \\ &\quad + \dot{\theta}^2 (I_z - I_y) S_\theta S_\varphi C_\varphi + 2\dot{\psi}\dot{\theta}\left\{C_\theta S_\theta\left(I_x - I_z C_\varphi^2 - I_y S_\varphi^2\right)\right\} \\ &\quad + 2\dot{\psi}\dot{\varphi}(I_y - I_z) S_\varphi C_\varphi C_\theta^2 - 2\dot{\varphi}\dot{\theta} C_\theta \left\{S_\varphi^2 (I_z - I_y) + \tfrac{1}{2}(I_x + I_y - I_z)\right\} \\ &\quad + Mh\left\{\ddot{X}_0 (C_\psi S_\varphi - S_\psi S_\theta C_\varphi) + \ddot{Y}_0 (S_\psi S_\varphi + C_\psi C_\varphi S_\theta)\right\} = Q_\psi,\end{aligned} \tag{5.42a}$$

$$\begin{aligned}&\ddot{\varphi} I_x - \ddot{\psi} S_\theta I_x + \dot{\psi}^2 (I_z - I_y) C_\theta^2 S_\varphi C_\varphi + \dot{\theta}^2 (I_y - I_z) S_\varphi C_\varphi \\ &\quad - \dot{\psi}\dot{\theta}\{I_x + (I_y - I_z)(C_\varphi^2 - S_\varphi^2)\} \\ &\quad + Mh\{\ddot{X}_0 (S_\psi C_\varphi - C_\psi S_\theta S_\varphi) - \ddot{Y}_0 (S_\psi S_\theta S_\varphi + C_\psi C_\varphi)\} + K_\varphi \varphi = Q_\varphi,\end{aligned} \tag{5.42b}$$

$$\begin{aligned}&\ddot{\theta}\left(C_\varphi^2 I_y + S_\varphi^2 I_z\right) + \ddot{\psi}(I_y - I_z) C_\theta C_\varphi S_\varphi + \dot{\psi}^2 C_\theta S_\theta \left(I_y S_\varphi^2 + I_z C_\varphi^2 - I_x\right) \\ &\quad + \dot{\psi}\dot{\varphi} C_\theta \left\{I_x + (I_y - I_z)(C_\varphi^2 - S_\varphi^2)\right\} + 2\dot{\varphi}\dot{\theta}(I_z - I_y) S_\varphi C_\varphi \\ &\quad + Mh(\ddot{X}_0 C_\psi C_\theta C_\varphi + \ddot{Y}_0 C_\theta S_\psi C_\varphi) + K_\theta \theta = Q_\theta,\end{aligned} \tag{5.42c}$$

$$\begin{aligned}&\bar{M}\ddot{X}_0 + Mh\{\ddot{\psi}(C_\psi S_\varphi - S_\psi S_\theta C_\varphi) + \ddot{\theta} C_\psi C_\theta C_\varphi + \ddot{\varphi}(S_\psi C_\varphi - C_\psi S_\theta S_\varphi) \\ &\quad - \dot{\psi}^2 (S_\psi S_\varphi + S_\theta C_\psi C_\varphi) - \dot{\theta}^2 C_\psi S_\theta C_\varphi - \dot{\varphi}^2 (S_\psi S_\varphi - C_\psi S_\theta C_\varphi) \\ &\quad - 2\dot{\psi}\dot{\theta} S_\psi C_\theta C_\varphi - 2\dot{\varphi}\dot{\theta} C_\psi C_\theta S_\varphi + 2\dot{\varphi}\dot{\psi}(C_\psi C_\varphi + S_\psi S_\theta S_\varphi)\} + K_x X_0 = f_x,\end{aligned} \tag{5.42d}$$

$$\begin{aligned}&\bar{M}\ddot{Y}_0 + Mh\{\ddot{\psi}(S_\psi S_\varphi + C_\psi S_\theta C_\varphi) + \ddot{\theta} S_\psi C_\theta C_\varphi - \ddot{\varphi}(C_\psi C_\varphi + S_\psi S_\theta S_\varphi) \\ &\quad + \dot{\psi}^2 (C_\psi S_\varphi - S_\psi S_\theta C_\varphi) - \dot{\theta}^2 S_\psi S_\theta C_\varphi - \dot{\varphi}^2 (S_\psi S_\theta C_\varphi - C_\psi S_\varphi) \\ &\quad + 2\dot{\psi}\dot{\theta} C_\psi C_\theta C_\varphi - 2\dot{\varphi}\dot{\theta} S_\psi C_\theta S_\varphi + 2\dot{\varphi}\dot{\psi}(S_\psi C_\varphi - C_\psi S_\theta S_\varphi)\} + K_y Y_0 = f_y,\end{aligned} \tag{5.42e}$$

where $\bar{M}$ is the total mass (i.e., it includes the mass of the drive system, the gyro disk, M, and the effective mass of the vibrating structure); S and C represent the sine and cosine of the angles indicated by their subscripts; and I_x, I_y, and I_z are the moments of inertia about the coordinate axes.

In order for these devices to be useful in practice, the oscillations θ and φ should be reasonably small. Furthermore, the coordinate ψ can be dropped as a generalized coordinate and the spin velocity, $\dot{\psi}$, may be assumed as a constant designated as Ω. Thus, the resulting equations are a set of coupled linear differential equations as shown below.

$$I_y\ddot{\theta} + (I_z - I_x)\Omega^2\theta + \Omega\dot{\varphi}(I_x + I_y - I_z) + K_\theta\theta + Mh(C_\psi\ddot{X}_0 + S_\psi\ddot{Y}_0) = 0, \tag{5.43}$$

$$I_x\ddot{\varphi} + (I_z - I_y)\Omega^2\varphi - \Omega\dot{\theta}(I_x + I_y - I_z) + K_\varphi\varphi + Mh(S_\psi\ddot{X}_0 - C_\psi\ddot{Y}_0) = 0, \tag{5.44}$$

$$\bar{M}\ddot{X}_0 + Mh\left\{C_\psi\ddot{\theta} + S_\psi\ddot{\varphi} - \Omega^2(S_\psi\varphi + C_\psi\theta) - 2S_\psi\Omega\dot{\theta} + 2C_\psi\Omega\dot{\varphi}\right\} + K_x X_0 = f_x, \tag{5.45}$$

$$\bar{M}\ddot{Y}_0 + Mh\left\{S_\psi\ddot{\theta} - C_\psi\ddot{\varphi} + \Omega^2(C_\psi\varphi - S_\psi\theta) + 2C_\psi\Omega\dot{\theta} + 2S_\psi\Omega\dot{\varphi}\right\} + K_y Y_0 = 0. \tag{5.46}$$

The above set of equations, although simplified considerably from its original form (5.42), cannot be solved easily because of the periodic coefficients $\sin\psi$ and $\cos\psi$. Further simplifications may, however, be accomplished by proper transformation of coordinates. Defining such a transformation as

$$\xi_1 = S_\psi \theta, \tag{5.47a}$$

$$\xi_2 = C_\psi \theta, \tag{5.47b}$$

$$\xi_3 = S_\psi \varphi, \tag{5.47c}$$

$$\xi_4 = C_\psi \varphi, \tag{5.47d}$$

and with some algebraic manipulation, a set of linear equations may be derived.

$$\ddot{\eta} + \Omega \frac{I_z}{I_x} \dot{\zeta} + \frac{Mh}{I_x} \ddot{X}_0 + \frac{K}{I_x} \eta = 0, \tag{5.48a}$$

$$\ddot{\zeta} - \Omega \frac{I_z}{I_x} \dot{\eta} + \frac{Mh}{I_x} \ddot{Y}_0 + \frac{K}{I_x} \zeta = 0, \tag{5.48b}$$

$$\ddot{\eta} + \frac{\bar{M}}{Mh} \ddot{X}_0 + \frac{K_x}{Mh} X_0 = \frac{f_x}{Mh}, \tag{5.48c}$$

$$\ddot{\zeta} + \frac{\bar{M}}{Mh} \ddot{Y}_0 + \frac{K_y}{Mh} Y_0 = 0, \tag{5.48d}$$

where

$$\eta = \xi_2 + \xi_3, \tag{5.49}$$

$$\zeta = \xi_1 - \xi_4. \tag{5.50}$$

Under steady-state conditions, the solutions for η, ζ, X_0, and Y_0 may be assumed as

$$\eta = \eta_0 e^{i\omega t}, \tag{5.51a}$$

$$\zeta = \zeta_0 e^{i\omega t}, \tag{5.51b}$$

$$X_0 = X_{0_1} e^{i\omega t}, \tag{5.51c}$$

$$Y_0 = Y_{0_1} e^{i\omega t}. \tag{5.51d}$$

Substitution of the assumed form of solution in the set of equations (5.48) yields the matrix equation

$$\begin{bmatrix} \frac{K}{I_x} - \omega^2 & i\frac{I_z}{I_x}\Omega\omega & -\frac{Mh\omega^2}{I_x} & 0 \\ -i\frac{I_z}{I_x}\Omega\omega & \frac{K}{I_x} - \omega^2 & 0 & -\frac{Mh\omega^2}{I_x} \\ -\omega^2 & 0 & \frac{K_x - \bar{M}\omega^2}{Mh} & 0 \\ 0 & -\omega^2 & 0 & \frac{K_y - \bar{M}\omega^2}{Mh} \end{bmatrix} \begin{Bmatrix} \eta \\ \zeta \\ X_0 \\ Y_0 \end{Bmatrix} = \begin{Bmatrix} 0 \\ 0 \\ \frac{f_x}{Mh} \\ 0 \end{Bmatrix}. \tag{5.52}$$

Assuming steady-state vibration, the displacements X_0 and Y_0 may be shown to be

$$\frac{DX_0}{f_0} = \omega^6 \left(\frac{Mh}{I_x} - \frac{\bar{M}}{Mh} \right) - \omega^4 \left(\frac{KMh}{I_x^2} - 2\frac{K}{I_x}\frac{\bar{M}}{Mh} - \Omega^2 \frac{I_z^2}{I_x^2}\frac{\bar{M}}{Mh} - \frac{K_y}{Mh} \right)$$
$$- \omega^2 \left\{ \frac{K^2}{I_x^2}\frac{\bar{M}}{Mh} + \frac{K_y}{Mh}\left(\omega^2 \frac{I_z^2}{I_x^2} + 2\frac{K}{I_x} \right) \right\} + \frac{K^2}{I_x^2}\frac{K_y}{Mh} \tag{5.53}$$

and

$$\frac{DY_0}{f_0} = \Omega \frac{I_z}{I_x^2} Mh\omega^5, \tag{5.54}$$

where D is the determinant of the coefficient matrix.

It may be noted from eq. (5.54) that the displacement along the Y direction can never be zero. Therefore, even though a null is attained along the X direction, the system may have oscillations in the orthogonal direction. This imposes a limitation on the usefulness of the perissogyro device. However, it is clear from eq. (5.54) that the displacement Y_0 is linearly related to the speed of the gyro wheel. This property will be used later to discuss the development of the so-called double perisso vibration absorber.

Equation (5.53) when equated to zero represents the null equation for the system under consideration. Equations (5.53) and (5.54) are quite general in that they include the effects of spring rates in the pivots as well as along the X_f, Y_f directions. General solutions for these equations may be obtained numerically. However, they can be simplified further if the spring rates are assumed as zero and if $\bar{M} = M$.

Thus, when $K = 0 = K_x = K_y$ and $\bar{M} = M$ the equation for the null frequencies is given by

$$\omega_n^2 = \frac{\Omega^2 I_z^2}{I_x(I_x - Mh^2)} \tag{5.55}$$

and the equation for the resonant frequencies may be shown to reduce to

$$\omega_r^2 = \frac{\Omega^2 I_z^2}{(I_x - Mh^2)^2}. \tag{5.56}$$

Thus,

$$\left(\frac{\omega_n}{\omega_r}\right)^2 = 1 - \frac{Mh^2}{I_x}; \tag{5.57}$$

that is,

$$\omega_n < \omega_r. \tag{5.58}$$

An examination of eq. (5.55) indicates an interesting contrast with conventional absorbers in that the null frequency depends on the magnitude of the absorber mass.

Equations (5.53) and (5.55) indicate that the null frequency is related to the speed of the gyro wheel. Such relationships are characteristic of gyroscopic systems and provide the unique advantage by means of which the absorbers may be synchronized, leading to antiresonance of the structure to which it is attached at all values of the driving frequency.

Numerical results obtained by solving the null and the characteristic equations for the single perissogyro are presented graphically in a later section. Comparison has been made with the results obtained by experiment.

The preceding analysis has shown that although the perissogyro is capable of producing antiresonance in the direction of the forcing function, oscillations will always occur in the orthogonal direction. These oscillations may be of relatively small amplitude, but nevertheless they may be undesirable. As suggested earlier, the vibrations induced in the Y direction are linearly related to the gyro angular velocity Ω. This leads one to believe that by superposing the effects of two perissogyro vibration absorbers, the gyro angular velocities of which are opposite to one another ($\Omega_1 = -\Omega_2$), the effect in the Y direction may altogether be eliminated. Such a device is designated here as the "double perisso" and consists of a set of two perissogyro vibration absorbers attached to the vibrating structure. The gyro wheel in one of the sets rotates in a direction opposite to that of the other.

5.4.2 Experimental Setup and Observations

An absorber was designed by using a Hooke's joint; one end was connected to a synchronous motor and the other to a circular aluminum plate at one end of a steel rod. Details of the device, a perissogyro, are listed in Table 5.1. The schematic of the setup is shown in Fig. 5.13. The speed and the direction of rotation of the motor could be controlled at either 1800 RPM or 3600 RPM. The entire absorber was then mounted on two shafts that rotated in opposite directions in order to remove the effect of friction forces along the direction of oscillation. Plunger springs were mounted laterally to provide a lateral spring rate, which was measured to be around 600 lb/in.

The antiresonant frequency was noted as follows: Setting the excitation frequency at a particular value, the motor turning at 3600 RPM was turned off. As the motor decelerated, the actual instantaneous RPM was read on an electronic counter when the null was located on the oscilloscope. The procedure was repeated for various values of the excitation frequency; the results are shown in Fig. 5.14 along with results from analysis. A comparison of the null-RPM characteristics predicted by theory (with K_y as parameter) with those obtained from experiment shows that the experimental values

Table 5.1. Experimental Model of the Single Perissogyro Absorber

Gyro disk material	Aluminum
Diameter of the disk	8 in.
Thickness of the disk	0.5 in.
Weight of gyro disk	2.44 lb
Length of steel rod	5.5 in.
Weight of motor assembly	24.2 lb
Weight of steel rod	0.625 lb
Weight of universal joint	0.24 lb
Weight of top plate	1.45 lb
Weight of connecting rod	1.50 lb
Weight of pick-ups	1.60 lb
Weight of shaker armature	2.00 lb

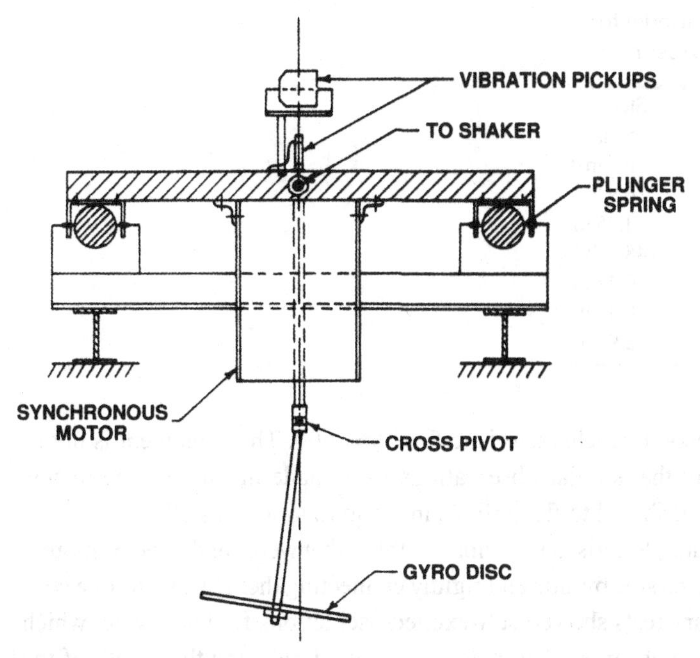

Figure 5.13. Schematic of the experimental setup of the perissogyro vibration absorber (Srinivasan, 1968).

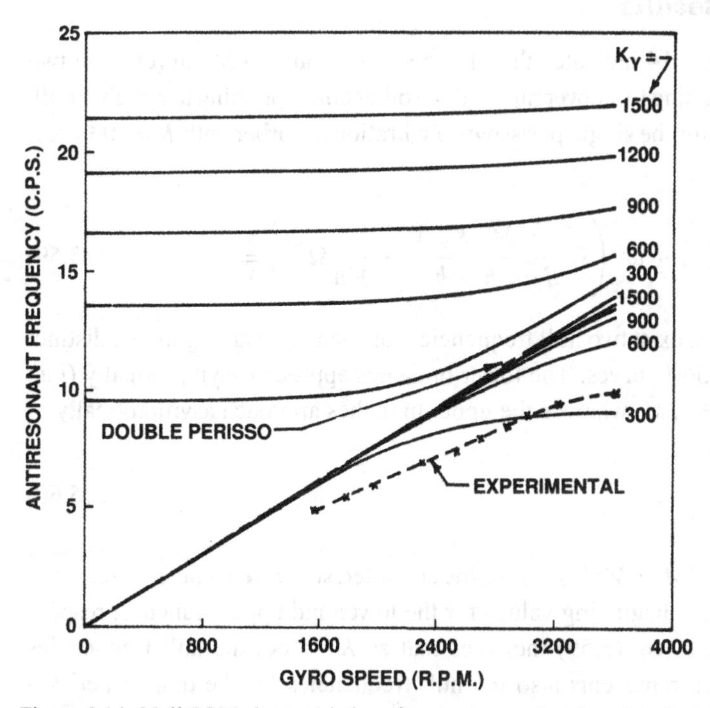

Figure 5.14. Null-RPM characteristics of perissogyro vibration absorbers (Srinivasan, 1968).

Table 5.2. Experimental Model for the Double Perissogyro Absorber

Gyro disk material	Steel
Diameter of the disk	8 in.
Thickness of the disk	0.5 in.
Weight of gyro disk	6.9 lb
Length of steel rod	3.75 in.
Weight of motor assembly	48.4 lb
Weight of two steel rods	0.72 lb
Weight of two universals	0.48 lb
Weight of top plate	2.90 lb

appear to parallel the theoretical characteristic for $K_y = 600$. The agreement is not as close as expected except that similar observations were made in earlier experiments made by Flannelly and attributed to flexibilities in the gyro system itself.

Experiments on the double perissogyro vibration absorber were conducted by mounting two single perissogyros side by side and rigidly connecting them by means of a connecting plate. Preliminary tests showed self-excited oscillations for the device, which required a redesign of the absorber. The redesign required reducing the length of the shaft to 3.75 in. and replacing the aluminum disk with a steel disk. Other details of the double perissogyro are shown in Table 5.2.

5.5 Numerical Results

An examination of Fig. 5.14 indicates that the theoretical null-RPM curves have two branches, one corresponding to a lower null and the other corresponding to a higher null. From the null equation for the single perissogyro vibration absorber with $K = 0 = K_x$, i.e.,

$$\omega^4\left(\frac{Mh}{I_x} - \frac{\bar{M}}{Mh}\right) + \omega^2\left(\Omega^2\frac{I_z^2}{I_x^2}\frac{\bar{M}}{Mh}\frac{K_y}{Mh}\right) - \frac{K_y}{Mh}\Omega^2\frac{I_z^2}{I_x^2} = 0, \tag{5.59}$$

it can be shown that there exist two null frequencies for each Ω, resulting in two distinct branches for the null-RPM curves. The lower branches approach asymptotically (i.e., as $\Omega \to \infty$) the value $K_y/\bar{M}$, whereas the upper branches approach asymptotically

$$\frac{\bar{M} I_z^2 \Omega^2}{I_x(\bar{M} I_x - M^2h^2)} \tag{5.60}$$

Also $\omega^2 = 0$, $\omega^2 = K_y/(\bar{M} - M^2h^2/I_x)$ are the so-called static frequencies (i.e., when $\Omega = 0$). The latter are the beginning values for the lower and upper branches, respectively. An examination of eq. (5.59) indicates that as $K \to \infty$, the null frequencies approach $\Omega I_z/I_x$, which represents also the null frequencies for the double perissogyro. This feature can be observed from Fig. 5.14, where the null-RPM curves for

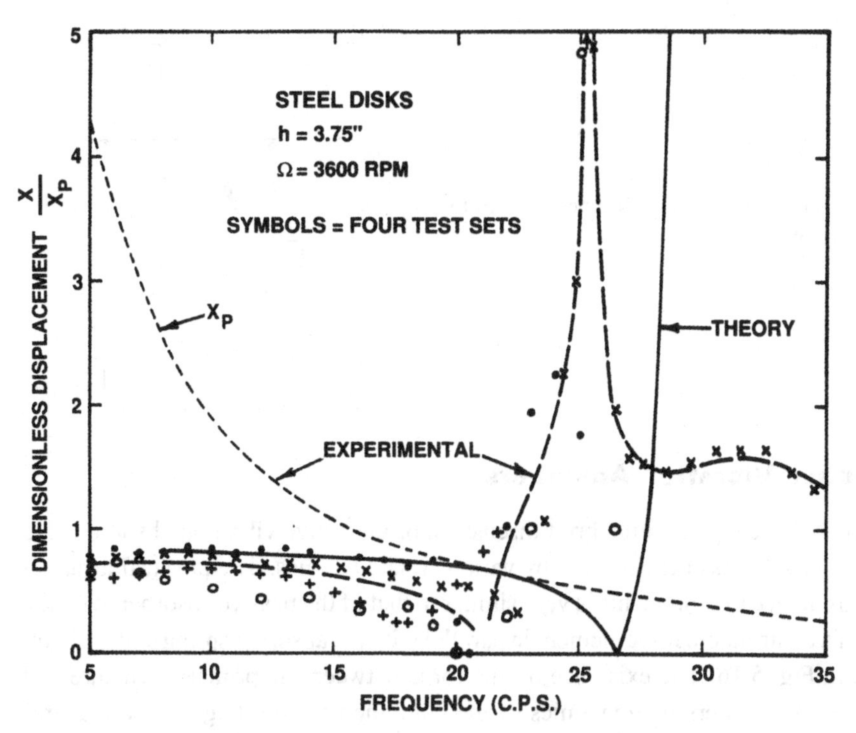

Figure 5.15. Theoretical and experimental response curves for the double perissogyro vibration absorber (Srinivasan, 1968).

the single perissogyro approach the corresponding curve for the double perissogyro as $K_y \to \infty$. This is appropriate because a double perissogyro always has a null in the lateral direction, a condition that corresponds to having an infinite lateral spring rate.

Figure 5.15 shows the principal features of the double perissogyro vibration absorber. Dimensionless values X/X_p are plotted against the forcing frequencies and compared with corresponding values obtained by experiment. Unlike the single perissogyro, the double perissogyro provides a single null. The general trends in these experimental observations conform closely to those suggested by theory.

One of the unique characteristics of a gyroscopic vibration absorber is the synchronization possibilities it provides. For the single perissogyro, the null-RPM characteristics are linear only over a limited range of gyro speeds, as may be seen from Fig. 5.14. For the double perissogyro, however, the null-RPM relationship is linear for any range of gyro speeds. Thus, a double perissogyro vibration absorber lends itself to linear synchronization, in addition to providing nulls in orthogonal directions.

Synchronous vibration absorbers hold promise as useful devices that may be incorporated into the original design of a structure to minimize vibration levels. By proper analysis of a structure, the effective masses at various stations may be determined so that the ideal position to locate the synchronous absorbers may be determined in the early design stages. Thus, the engineer can define the most efficient structure for strength and weight and reduce vibration problems to a desired minimum.

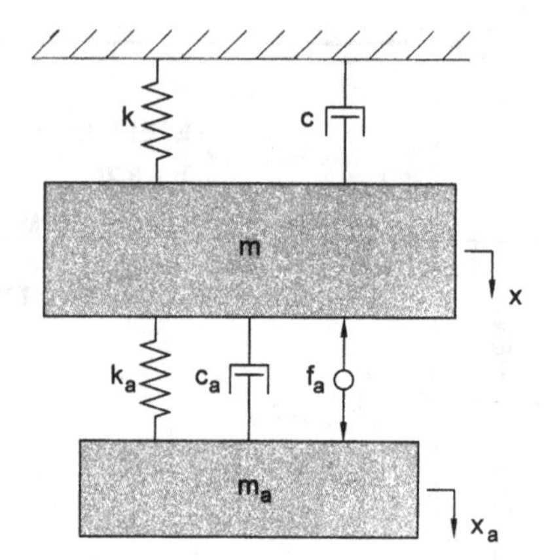

Figure 5.16. An active vibration absorber attached to a primary structure.

5.6 Active Vibration Absorbers

We conclude this chapter with a brief discussion of the active vibration absorber, also known as a proof-mass actuator and by various related names. The physical arrangement of this device's components is very similar to that of the passive absorber, with the addition of a controllable force element in parallel with the absorber spring and dashpot, as shown in Fig. 5.16. The existence of this force between the primary structure and the absorber mass alters the dynamics of the combined system (e.g., its natural frequencies and the magnitude of the response of the main mass to a harmonic disturbing force).

In many applications the active control force f_a is derived by feeding back a combination of the relative displacement and velocity between the primary and secondary masses. This results in a system that is mathematically equivalent to a passive absorber in which the spring and dashpot are adjustable, allowing the absorber to be tuned during operation. Although such a system may perform better in the harmonic steady state than a conventional, passive device, with the introduction of active feedback comes the potential for instability and other undesirable phenomena to which passive structures are generally immune (see Appendix B and the exercises at the end of Chapter 8).

In addition, the generation of the force f_a, for example by a hydraulic actuator, consumes power, and the associated hardware increases the complexity of what is basically a simple device. These drawbacks must be weighed against the benefits that may follow from flexibility in specifying f_a in a given application. For an example of the integration of proof-mass actuators into a complex structural system see Wie (1988).

5.7 Summary

We have examined a few devices that may be developed further in order to attain a design objective of enhanced bandwidth within which vibratory energy may be absorbed. These schemes include an open-loop concept (parallel damped dynamic vibration absorber) and several closed-loop concepts. The applicability of these devices depends upon the requirements to be met in vibration attenuation. With the vast improvements that have taken place in electronics technology, gyroscopic absorbers may find increased areas of application. Parallel vibration absorbers are simple devices that may be better in certain applications than the conventional Frahm absorber.

BIBLIOGRAPHY

Den Hartog, J. P. 1985. *Mechanical Vibrations*. 4th ed. New York: Dover Publications, Inc.

Flannelly, W. G., and J. C. Wilson. 1967. Analytical research on a synchronous gyroscopic vibration absorber. Technical Report CR-338, NASA.

Srinivasan, A. V. 1968. Analytical and experimental studies of gyroscopic vibration absorbers (Part I) and analysis of parallel damped vibration absorbers (Part II). Technical Report NASw-1394, NASA.

Wie, B. 1988. Active vibration control synthesis for the control of flexible structures mast flight system. *AIAA Journal of Guidance, Control and Dynamics* 11(3):271–277.

PROBLEMS

1. Derive eq. (5.7).

2. Identify application areas where a null amplitude at an antiresonance is desirable. Similarly identify structural systems in which the operating range may include varying forcing frequencies.

3. Derive eqs. (5.42) and (5.48).

4. Discuss the implications of the assumptions made in obtaining eqs. (5.55) and (5.56).

5. A single-degree-of-freedom structure consists of a mass of 50 kg supported by a spring of stiffness 12 500 N/m and is viscously damped with damping ratio $\zeta = 0.7\%$. The mass is disturbed by the harmonic load $f(t) = f_0 \sin \omega t$.

(a) Find the frequency response function of this system, that is, $H(j\omega) = X/F$, where $x(t)$ is the displacement of the mass.

(b) A viscously damped SDOF vibration absorber is now attached to the main mass of the given system. Calculate $H(j\omega)$, taking into account the effects of this additional degree of freedom.

(c) Plot the maximum response amplitude against absorber damping ratio and mass ratio.

6. Repeat the analysis of Problem 5(b) for an absorber with both damped and undamped degrees of freedom, with masses m_2 and m_3, respectively. For a fixed absorber damping coefficient and total absorber mass $m_2 + m_3$, examine the effects of dividing the absorber mass between m_2 and m_3 in various ratios.

7. For the experimental configuration shown in Fig. 5.13, calculate the displacement in the lateral direction at the theoretical null frequency if no plunger springs are provided. Discuss the advantages/disadvantages of using plunger springs compared to a "double perisso" device.

8. A gyroscopic vibration absorber is to be designed to suppress the motion of the primary structure described in Problem 5.

(a) Using the simplest applicable method of analysis from this chapter, determine the size of a GVA disk of mass 2 kg.

(b) Find the frequency range over which the GVA from part (a) is effectual if its rotational speed Ω is limited to a maximum of 3600 RPM.

6

Mistuning

6.1 Introduction

Mistuned structural systems represent periodic structures (structures with repeating segments, such as a multi-span beam of nominally equal spans) in which one or more physical or dynamic characteristics (for example, length, mass, material properties, frequency, or damping) vary from sector to sector. A tuned system, therefore, represents a structure in which the physical parameters are identical in each segment. The discrepancies or departures in these parameters may be only slight perturbations from those of the corresponding tuned system. Nevertheless, the modal characteristics of the perturbed system may bear little or no resemblance to modes of the tuned system. Exploiting the dynamics of the structure by (a) taking advantage of the nonuniformities present or (b) deliberately designing nonuniformities into the structure (mistuning), in order to meet or exceed the design intent, constitutes open-loop smartness. An example of a design intent is avoiding instabilities or reducing resonant stresses.

A mistuned structure is shown in Fig. 6.1, which represents a two-span beam in which the spans are nominally identical but in fact the length of one span is slightly different from the length of the other. Another example, which we shall discuss in some detail, is a turbine disk on which a hundred or so blades are mounted (Fig. 6.2), each of which has a fundamental frequency slightly different from that of its neighbors. Such a difference in frequency may arise between blades supplied by different vendors, manufacturing tolerances notwithstanding.

From the two examples given above, it should be clear that mistuned systems are more a rule than an exception in engineering practice. In the turbine wheel example, even if the blades were identical to each other to start with, they would most likely not wear uniformly in service, resulting in a mistuned system. Also in the case of turbines, it is important to note that mistuning may arise because of slightly different aerodynamics from passage to passage. Thus, in general, mistuning may arise from nonuniformities in material, structural, or aerodynamic characteristics.

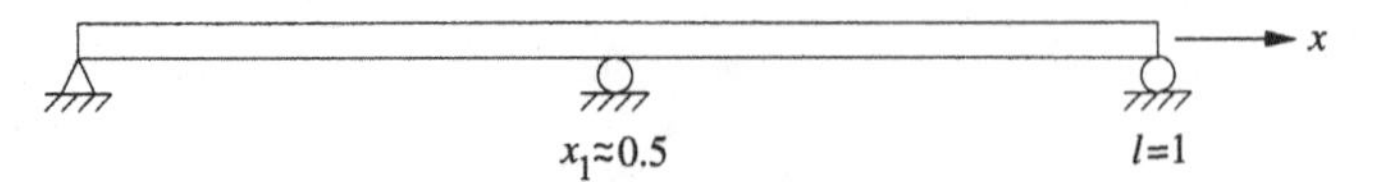

Figure 6.1. Two-span simply supported beam.

6.2 Vibration Characteristics of Mistuned Systems

We shall use the same examples to further describe the underlying physical mechanisms and modal characteristics of mistuned systems.

6.2.1 Nearly Periodic Simply Supported Beams

The uniform two-span beam of length $l = 1$ shown in Fig. 6.1 is tuned if and only if the interior support is located in the center of the beam (i.e., $x_1 = 0.5$); otherwise it is mistuned or disordered. We shall examine the effect on the natural frequencies and modes of this system of moving the central support slightly off-center. Details of the analysis may be found in Bergman and McFarland (1988).

Pierre, Tang, and Dowell (1986) considered the localization of free vibration modes produced by mistuning. Localization is a phenomenon wherein the amplitude of vibration of one span is greater than that of the others, potentially dominating the structure's response in that mode. They defined a measure of localization, denoted A, as the ratio of the maximum deflection in a span to the maximum deflection in the entire beam, so that $0 \leq A \leq 1$. The smaller the value of A, the greater the degree of localization of the corresponding mode; for a tuned structure, $A = 1$ for all modes.

The first six modes of the two-span beam are shown in Fig. 6.3 for tuned ($x_1 = 0.50$) and mistuned ($x_1 = 0.51$) configurations. The modes of the latter exhibit clearly visible localization. Note that, depending upon which mode is examined, the greater

Figure 6.2. A first-stage turbine rotor (Srinivasan, 1997).

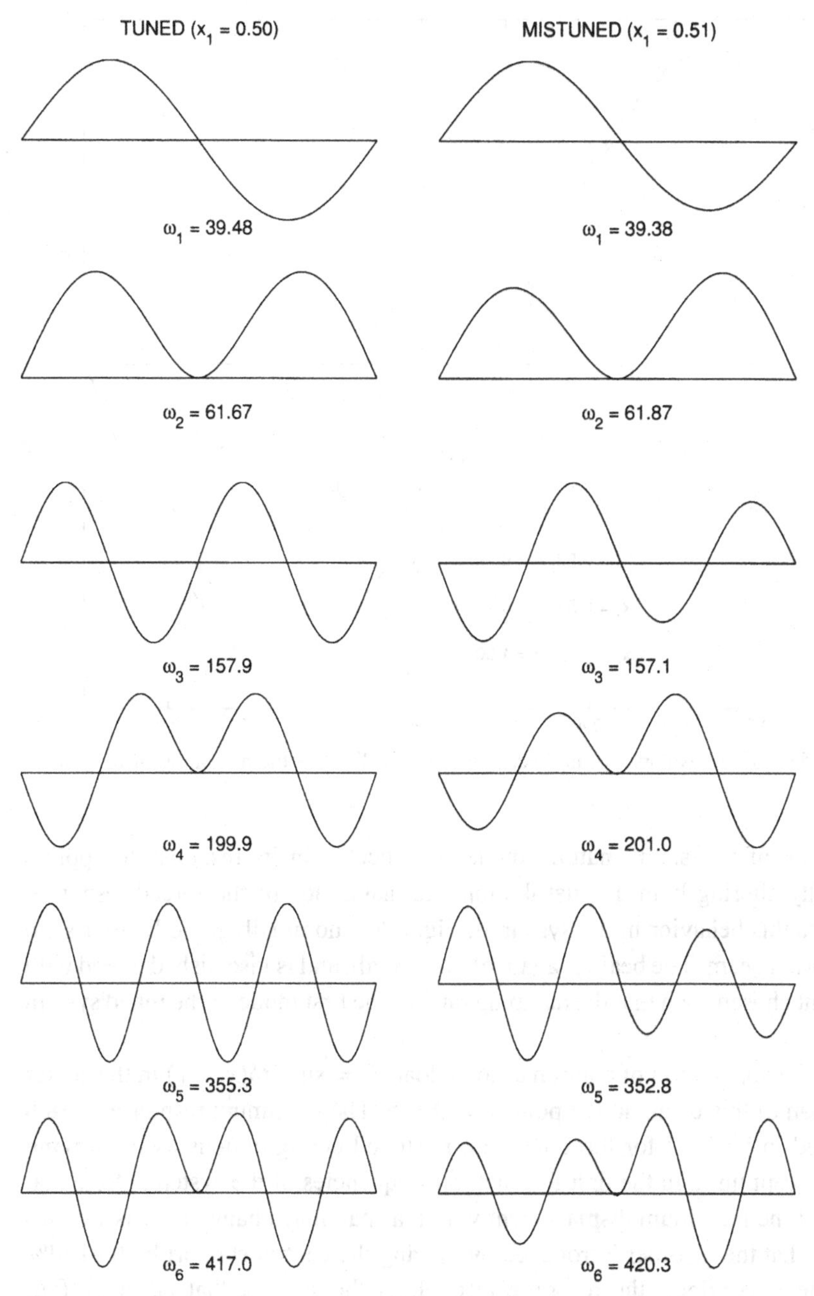

Figure 6.3. Natural frequencies and modes of the two-span beam, tuned and mistuned.

displacement can occur in either the longer or the shorter span. Figure 6.4 illustrates the effect on the shape of the first natural mode of increasing the mistuning by moving the middle support to the right. The relationship between the amount of mistuning and localization of the first mode is plotted in Fig. 6.5, where the localization measure A may be seen to decrease rapidly with increasing structural asymmetry.

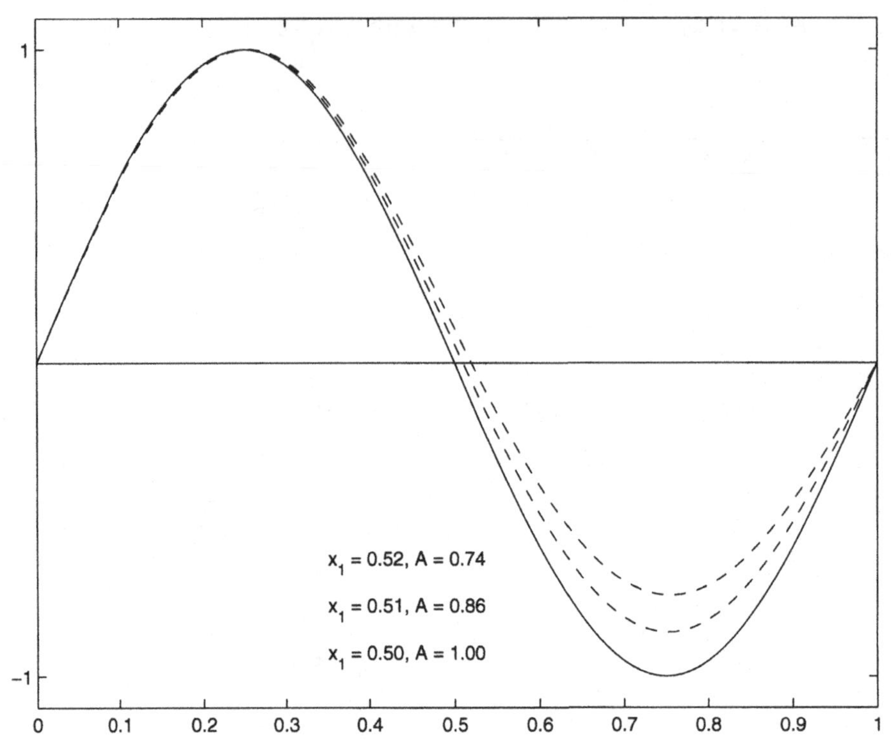

Figure 6.4. Mode 1 of the two-span beam becomes more localized as the mistuning of the system increases.

Localization of a system's natural modes is reflected in its response to applied loads, typically altering both the distribution and magnitude of the forced response. We investigate this behavior in the system of Fig. 6.6, a nominally periodic four-span simply supported beam. The beam is again of unit length, and is viscously damped with this coefficient chosen such that the damping ratio of the first mode of the tuned system is 2%.

Consider the application of a unit harmonic load $P = \sin \omega t \delta(x - \hat{x})$ in the center of the first span of this beam, at the point $\hat{x} = 0.125$. The maximum response in each span is plotted in Fig. 6.7 for the tuned and mistuned configurations as the driving frequency is swept through the first four natural frequencies of the system. (Note that the location of the maximum displacement within a span may change as ω is varied.) It is apparent that the disorder introduced by moving the central support by 0.01 (4% of the span length) affects the response throughout the system, that is, in all four spans.

Next, this harmonic load is replaced by a random force modeled as delta-correlated white noise, $R_{PP} = \delta(x - \hat{x})\delta(\tau)$, again acting at the center of the leftmost span. As before, mistuning is introduced by relocating the middle support slightly off center, and the response of the beam is computed by modal analysis (Bergman and McFarland, 1989). The mean square displacement response of the beam is plotted over its length in Fig. 6.8. Although the response of the tuned structure is nearly symmetric about the

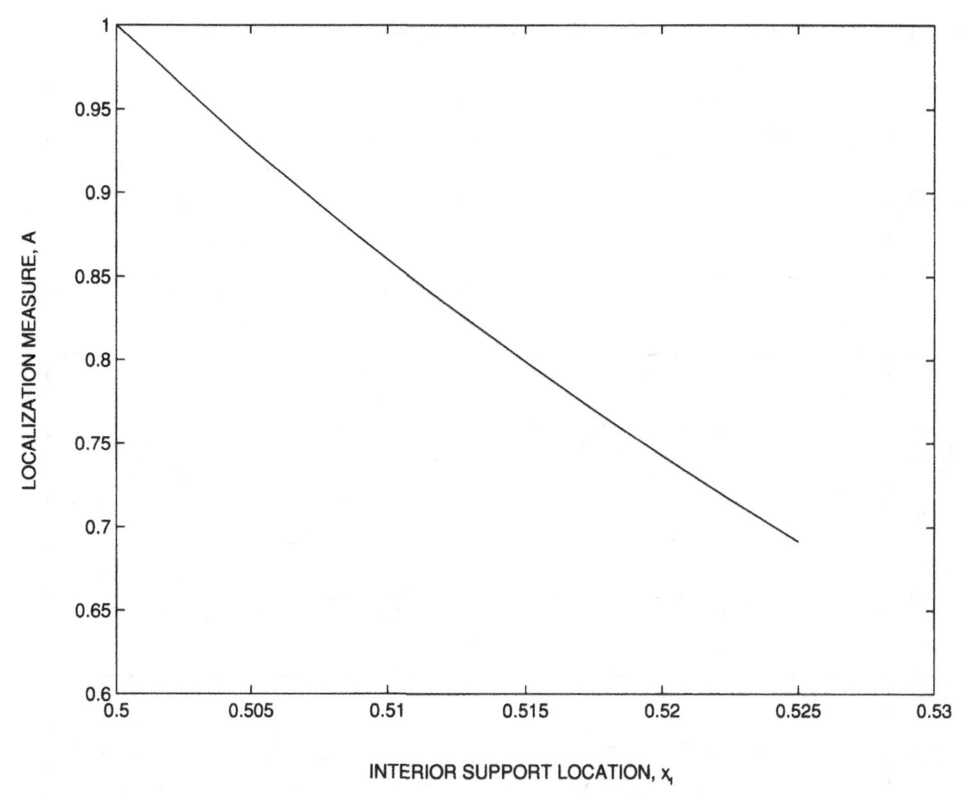

Figure 6.5. Localization of the first natural mode of the two-span beam grows (A decreases) as the interior support is moved away from the center of the beam.

midpoint of the beam, as the center support is moved the response becomes markedly less symmetric. In general, the magnitude of the response in the third and fourth spans is decreased at the expense of a pronounced increase in the displacement of the second span and a much smaller increase in the displacement of the first.

This localization of the forced response tends to confine motion, and thus energy, stress, and fatigue damage, to the region between the driven point and the mistuned span. This confinement is also observed in structures with much larger numbers of nominally periodic substructures. Exploiting this phenomenon is often a convenient way to introduce open-loop smartness into a structure, but it requires awareness of the trade-offs involved in each application.

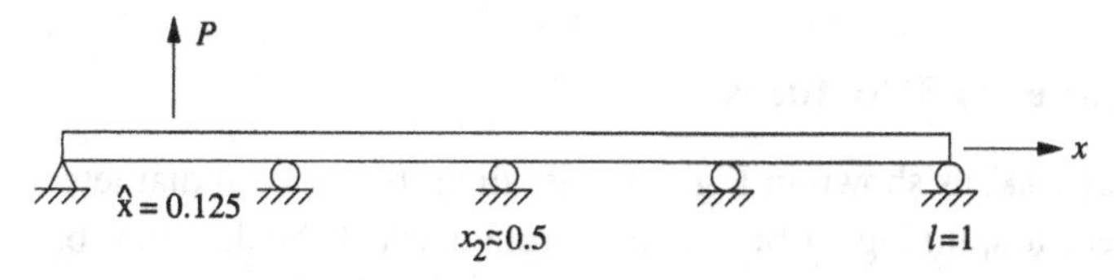

Figure 6.6. Point loaded four-span beam, mistuned by varying the location x_2 of the center support.

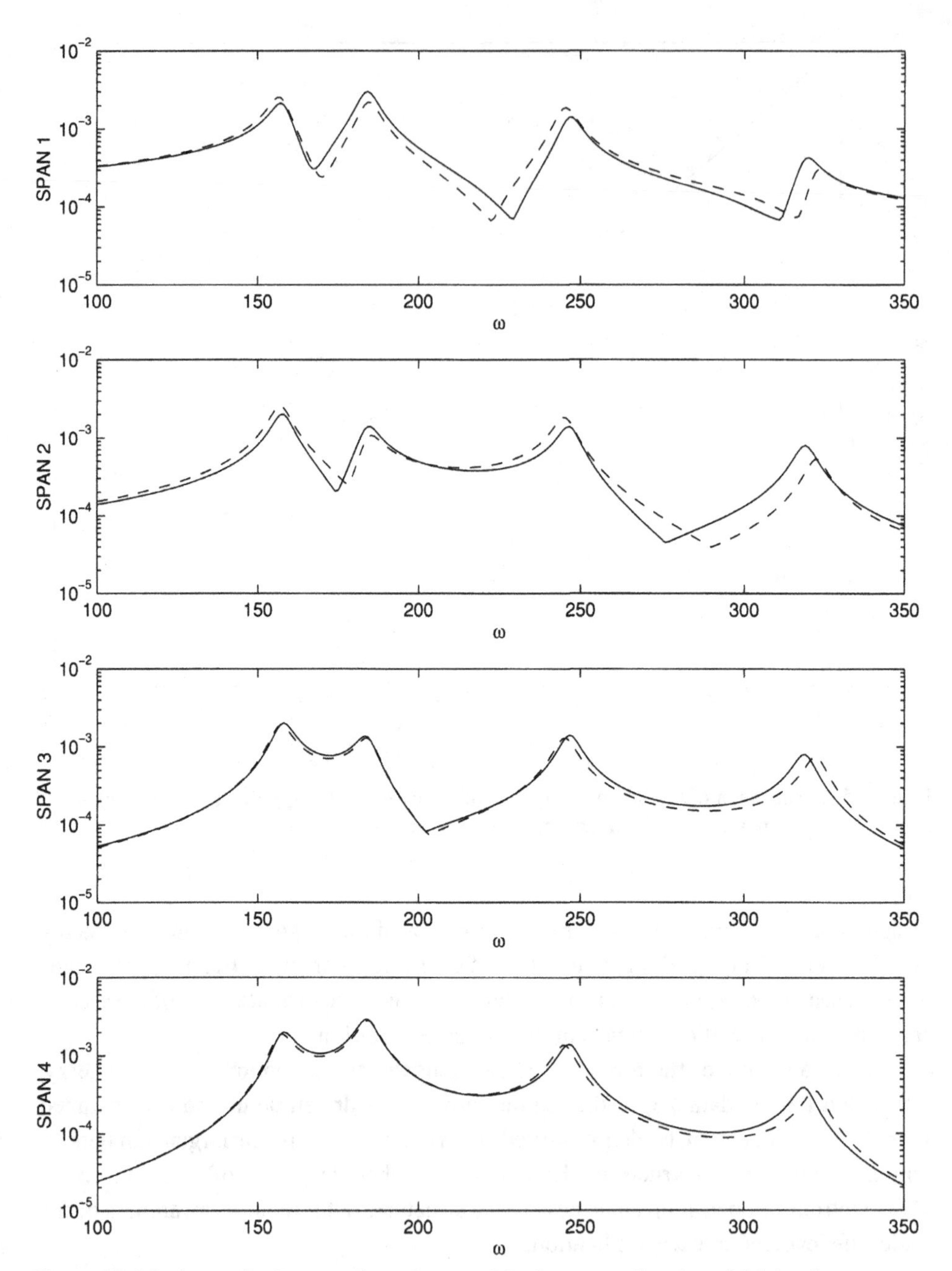

Figure 6.7. Maximum displacement in each span of the beam as a function of driving frequency ω. Solid line, tuned ($x_2 = 0.500$); broken line, mistuned ($x_2 = 0.51$).

6.2.2 Circularly Symmetric Structures

Consider a circular plate or disk, as shown in Fig. 6.9, vibrating in a 2 nodal diameter pattern. This may represent a spinning turbine wheel around which blades may be mounted. When a forcing function acts on the plate in such a way that it matches the

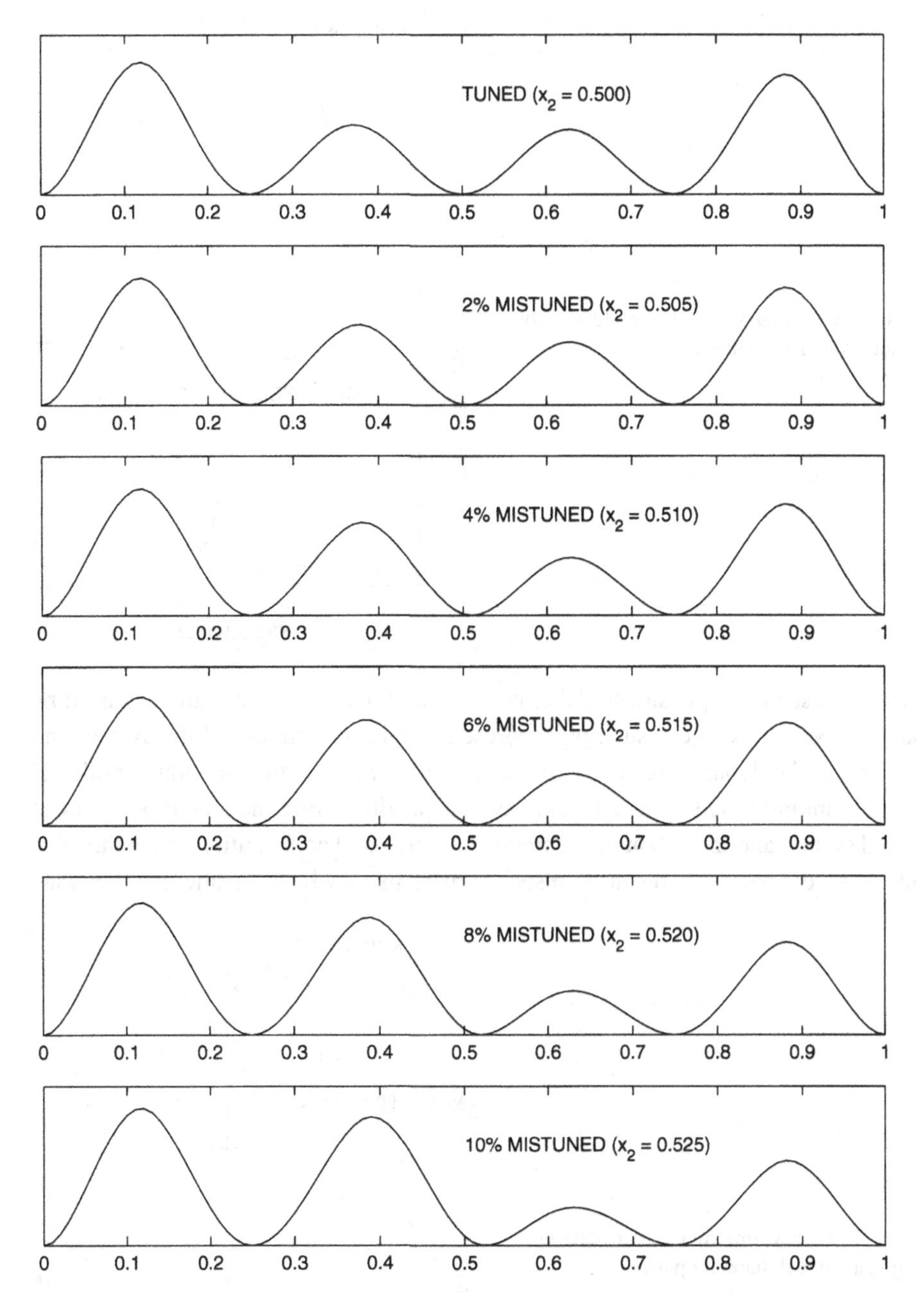

Figure 6.8. Mean square displacement of the four-span beam is increasingly localized as the system becomes more disordered.

natural mode both in space and time, then the vibratory amplitudes reach a peak at the natural frequency (ω_2) corresponding to the $n = 2$ mode. However, in the presence of any asymmetry, the well-defined $n = 2$ mode splits into two distinct but closely spaced frequencies (ω_{-2} and ω_{+2}) (Fig. 6.10). The spread between these frequencies depends upon the level of mistuning. The split modes are orthogonal to each other

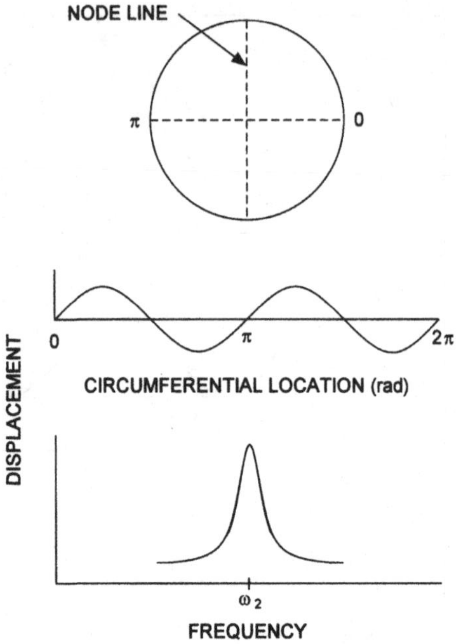

Figure 6.9. A circular symmetric plate vibrating in a 2 nodal diameter pattern.

and share the same simple sinusoidal circumferential distribution but are displaced by a quarter wave. Thus, every such pair represents legitimate modes of the system and contributes to the dynamic response of the structure, much as the individual modes of the corresponding tuned system would have. Depending upon the modal densities of the tuned system and the extent of the frequency splits and contributions from the wide variety of modal pairs, the resulting response patterns may be so erratic that they may

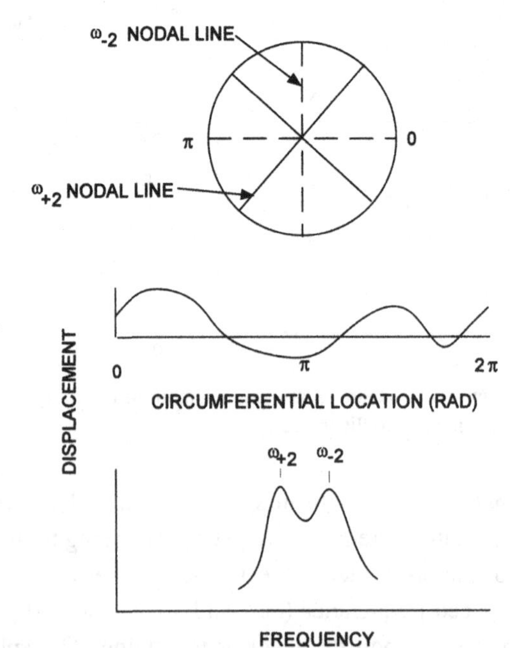

Figure 6.10. An asymmetric circular plate vibrating in a 2 nodal diameter pattern.

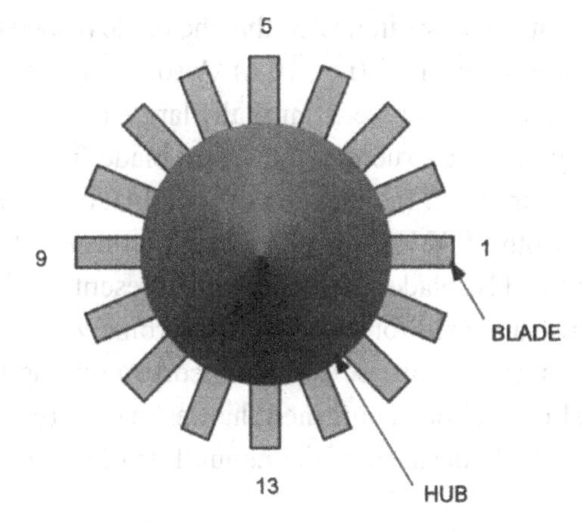

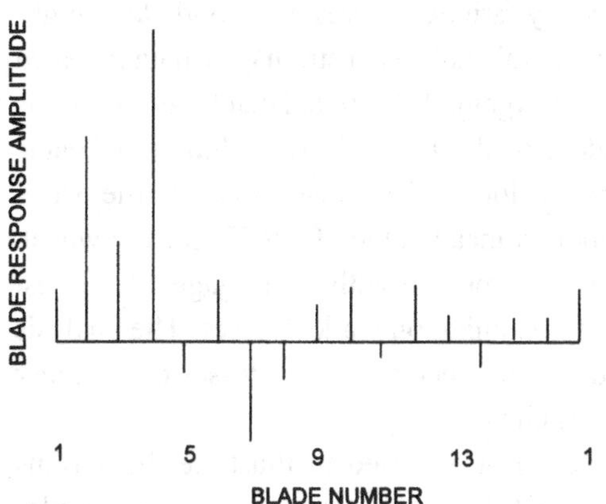

Figure 6.11. Amplitudes of vibration of a mistuned bladed-disk assembly.

defy a definition of a predominant harmonic. If blades are mounted on such a disk, then the amplitudes of vibration may have a pattern such as that shown in Fig. 6.11, with some blades experiencing disproportionately larger amplitudes than their neighbors. Thus, the amplitude of a blade depends not only on its own characteristics but also on its location around the rotor. In this figure, the lengths of bars sketched around the disk (0 to 2π) represent the amplitudes of motion at a natural frequency of the mistuned system. It is interesting to note parenthetically the difficult task faced by test engineers who have to determine the blades to be gaged in a rig or engine test program.

With this background, it is important to pose a question in the context of smart structures: Can the designer exploit the drastic changes in the dynamics of mistuned flexible structures to accomplish a design intent in a smart way? In order to underscore the need for considering this potential, we review below the field experience with gas turbine engines by briefly citing three incidents.

6.2.3 Jet Engine Blades

In two separate and distinct advanced engine designs, some turbine blades were found to experience high-frequency fatigue failures. In one case, failure occurred in the first-stage high-pressure turbine blades. Data analysis indicated that the blade response was due to a strong engine order excitation of order 10 (i.e., 10 E). Also, it was noted that vibratory stresses in some blades around the rotor were unusually large, indicating features uncharacteristic of a cyclically symmetric structure. Individual blade frequencies in their fundamental mode were measured and the location of each blade around the wheel was noted. A frequency bandwidth of 90 Hz was noted about a mean value of 1 343 Hz. Twenty-four blades out of the 116 blades on the rotor, representing the high, mean, and low frequencies, were instrumented for testing of the turbine wheel in a spin pit in which the assembly was subjected to air jet excitation corresponding to 10 E. Both measured data and analytical predictions confirmed that resonant stresses in certain blades can escalate as much as 95% because of the nonuniform frequency distribution.

In the second case, repeated high-frequency fatigue failures of second-stage high-pressure turbine blades were encountered. Spectral analysis of strain gage data revealed the response was due to a severe 5 E excitation. Again, it was noted that large variations in blade vibratory stresses existed from blade to blade around the rotor. Individual blade frequencies were measured and the attachment slot number of each blade in the wheel was noted. A scatter of 66 Hz was noted about a mean value of 876 Hz for the wheel. Twelve of the 72 blades on the rotor were instrumented with strain gages. Both test data and analysis confirmed the influence of mistuning on blade stresses. The analysis indicated that some blades on the rotor could experience resonant stresses 64% higher than they would in a corresponding tuned system.

Finally, the results from a compressor rig test are used to illustrate the serious influence of mistuning in some cases. Figure 6.12 shows three neighboring blades whose natural frequencies were about 9% and 24% higher than the lowest frequency, arranged in order of high, higher, and low frequency around the rotor. Under a forced aerodynamic excitation at 140 Hz, the neighboring blades experienced highly uneven stresses that varied as $10x$ and $4x$, where x is the lowest stress measured in the three

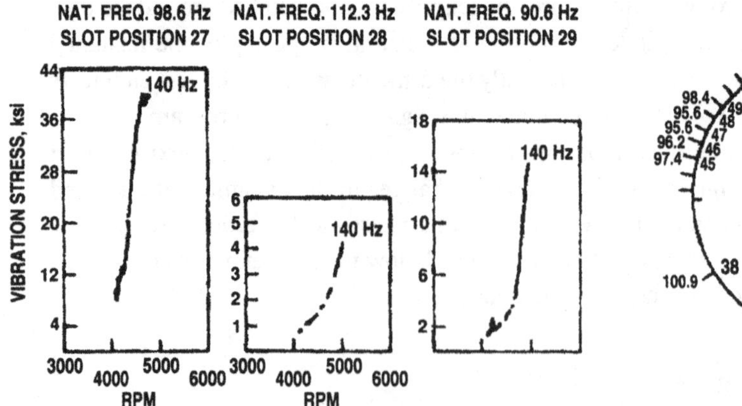

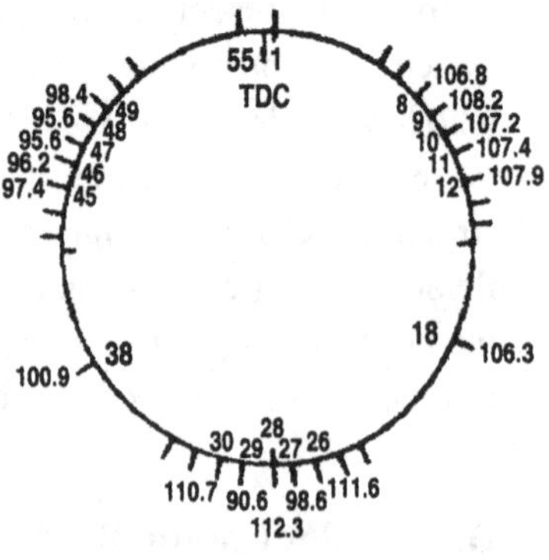

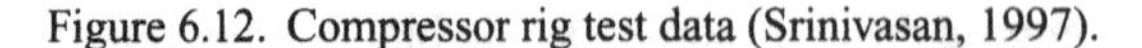
Figure 6.12. Compressor rig test data (Srinivasan, 1997).

blades. (It must be noted that the frequency spread shown above for the rig testing is not typical of compressor rotors.)

Mistuning is an important consideration in the design and development of turbine blades and has been a subject of intense research in the past four decades in industry, government, and academia. The research efforts were motivated by several incidents, such as the ones cited above in which unexpected and unexplained failure of blades in service in some engines occurred even though blades of the same design in other engines performed satisfactorily.

6.3 Analytical Approach

Let us now review an analytical procedure that may be used to evaluate the effects of mistuning on both blade resonance and flutter. Blade resonance refers to escalation of stresses in blades as the engine speed coincides with a natural frequency of the system. It is caused by unsteady aerodynamic forces' acting on a blade row. Flutter is aeroelastic instability caused by aerodynamic forces induced by blade vibration. Although resonant stresses de-escalate upon passing through a natural frequency, flutter stresses continue to escalate when once initiated and are controlled entirely by the damping available in the system.

With reference to Fig. 6.13, let us define

$$M_k = \text{mass of } k\text{th blade,} \tag{6.1a}$$

$$K_k = \text{stiffness of } k\text{th blade,} \tag{6.1b}$$

$$C_k = \text{material damping at } k\text{th blade,} \tag{6.1c}$$

$$A_k = \text{aeroelastic force on } k\text{th blade,} \tag{6.1d}$$

$$f_k = \text{force transmitted at blade/disk junction,} \tag{6.1e}$$

$$N = \text{number of blades,} \tag{6.1f}$$

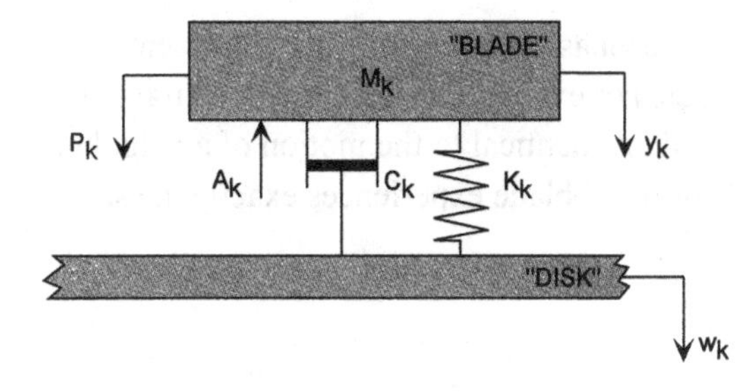

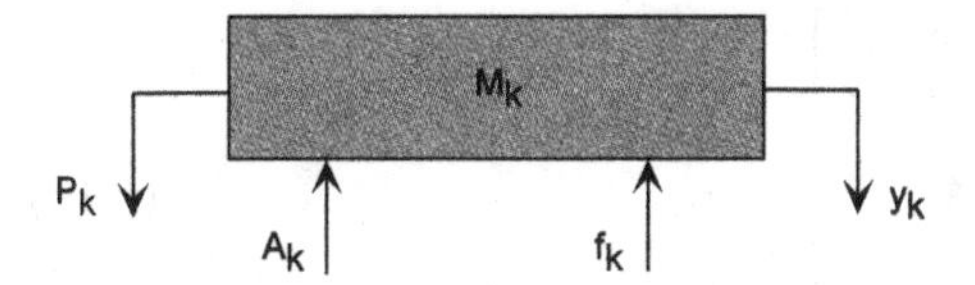

Figure 6.13. Aeroelastic model of a bladed-disk assembly.

$$P_k = \text{external force on } k\text{th blade,} \tag{6.1g}$$

$$y_k = \text{vibratory amplitude of } k\text{th blade,} \tag{6.1h}$$

$$w_k = \text{vibratory amplitude of disk at } k\text{th blade, and} \tag{6.1i}$$

$$z_k = y_k - w_k. \tag{6.1j}$$

Forces acting on each blade may be represented as the algebraic sum of inertial, elastic, aeroelastic, and mechanical damping forces in addition to the external load on each blade; that is,

$$M_k \ddot{y}_k + C_k \dot{z}_k + K_k z_k + A_k = P_k, \tag{6.2}$$

$$f_k = C_k z_k + K_k z_k, \tag{6.3}$$

$$M_k \ddot{y}_k + f_k + A_k = P_k. \tag{6.4}$$

Assume all quantities to have the common factor $e^{i\omega t}$, so that it can be factored out. Equation (6.2) may be written as

$$-\omega^2 M_k y_k + f_k + A_k = P_k, \tag{6.5}$$

where ω is the frequency of the forcing function. In matrix form,

$$-\omega^2 \lceil M \rfloor \{y\} + \{f\} + \{A\} = \{P\}. \tag{6.6}$$

Let u_r be y_k in the rth mode. The total amplitude of the kth blade is the sum of all complex amplitudes from all contributing modes, that is,

$$y_k = \sum_{r=1}^{N} u_r e^{ik\beta_r}, \tag{6.7}$$

where β is the interblade phase angle

$$\beta_r = \frac{2\pi r}{N} = r\beta_1. \tag{6.8}$$

Note $k\beta_r = r\beta_k$. (Interblade phase angle is a phase relationship that represents the motion of a blade with respect to other blades. For example, in a well-defined traveling wave mode of vibration, the motion of a blade is identical to the motion of a neighbor except for a phase angle between them. Thus, each blade experiences exactly the same amplitude, but at a different time.) Hence

$$\{y\} = [E]\{u\}, \tag{6.9}$$

where

$$[E] = \begin{bmatrix} e^{i\beta_1} & e^{i2\beta_1} & e^{i3\beta_1} & \dots & e^{iN\beta_1} \\ e^{i2\beta_1} & e^{i4\beta_1} & e^{i6\beta_1} & \dots & e^{i2N\beta_1} \\ e^{i3\beta_1} & \dots & \dots & \dots & e^{i3N\beta_1} \\ \vdots & & & & \vdots \\ e^{iN\beta_1} & & & & e^{iN^2\beta_1} \end{bmatrix}_{N\times N}. \tag{6.10}$$

The aeroelastic force on the kth blade due to motion in the rth mode alone is

$$A_{k_r} = 2\pi\rho UbC_{F_{ur}}\dot{u}_r e^{ik\beta_r} = 2\pi\rho Ubi\omega C_{F_{ur}} e^{ik\beta_r}, \tag{6.11}$$

where

$$b = \text{semichord}, \tag{6.12a}$$

$$U = \text{flow velocity}, \tag{6.12b}$$

$$\rho = \text{air density}, \tag{6.12c}$$

$$C = \text{unsteady aerodynamic force coefficient}. \tag{6.12d}$$

The total contribution from *all* modes to the total aeroelastic force on the kth blade is

$$A_k = 2\pi\rho Ubi\omega \sum u_r C_{F_{ur}} e^{ik\beta_r} = \varepsilon \sum u_r C_{F_{ur}} e^{ik\beta_r}, \tag{6.13}$$

where

$$\varepsilon = 2\pi\rho Ubi\omega, \tag{6.14}$$

or as a matrix equation,

$$\{A\} = \varepsilon[E]\lceil C\rfloor\{u\}. \tag{6.15}$$

Note $C_r = C(\lambda_k, \beta_r)$ where λ is reduced frequency $(b\omega/U)$.

Equation (6.6) may now be rewritten as

$$-\omega^2\lceil M\rfloor\{y\} + \{f\} + \varepsilon[E]\lceil C\rfloor\{u\} = \{P\}, \tag{6.16}$$

and upon substituting eq. (6.9),

$$-\omega^2\lceil M\rfloor\{y\} + \{f\} + \varepsilon[E]\lceil C\rfloor[E]^{-1}\{y\} = \{P\}. \tag{6.17}$$

From (6.3),

$$\{f\} = \lceil K_k + i\omega_k c_k\rfloor \{y - w\} \tag{6.18}$$

and therefore

$$\{y - w\} = \left[\frac{1}{K_k + i\omega_k C_k}\right]\{f\} = [\alpha_2]\{f\}. \tag{6.19}$$

The displacements around the disk rim at any blade location may be written as

$$w_k = \sum_{\ell=1}^{N} \alpha_{k\ell} f_\ell \tag{6.20}$$

or

$$\{w\} = [\alpha_1]\{f\}. \tag{6.21}$$

Therefore,

$$\{y\} = [\alpha_1 + \alpha_2]\{f\} \tag{6.22}$$

and

$$\{f\} = [\alpha_1 + \alpha_2]^{-1}\{y\}. \tag{6.23}$$

Substituting in (6.17), we get

$$[\varepsilon[E][C][E]^{-1} - \omega^2 M + [\alpha_1 + \alpha_2]^{-1}]\{y\} = \{P\}. \tag{6.24}$$

Equation (6.24) may be reduced (for $\{P\} = \{0\}$) to

$$[\mathcal{M} - \mathcal{A}]\{f\} = \nu\{f\}, \tag{6.25}$$

which is a complex eigenvalue problem where $\mathcal{M}$ includes mechanical effects and $\mathcal{A}$ includes aerodynamic influences.

Details of the elements of the matrices have been omitted because they can be quite complicated and our purpose here is to introduce the reader to the potentials of exploiting the dynamics of mistuned aeroelastic systems to design a smart structure. Interested readers may refer to Srinivasan (1997) to obtain an overview of this important subject. If the variable y is expressed as $y = ye^{\nu t}$ where ν is a measure of dimensionless frequency (a reference frequency may be the frequency of, say, blade number 1), then the real part of the eigenvalue is a measure of the vibratory modal frequency and the imaginary part is a measure of damping.

When the eigenvalue problem is solved for all possible interblade phase angles at a prescribed frequency, the solution can be plotted as a locus as shown in Fig. 6.14, from

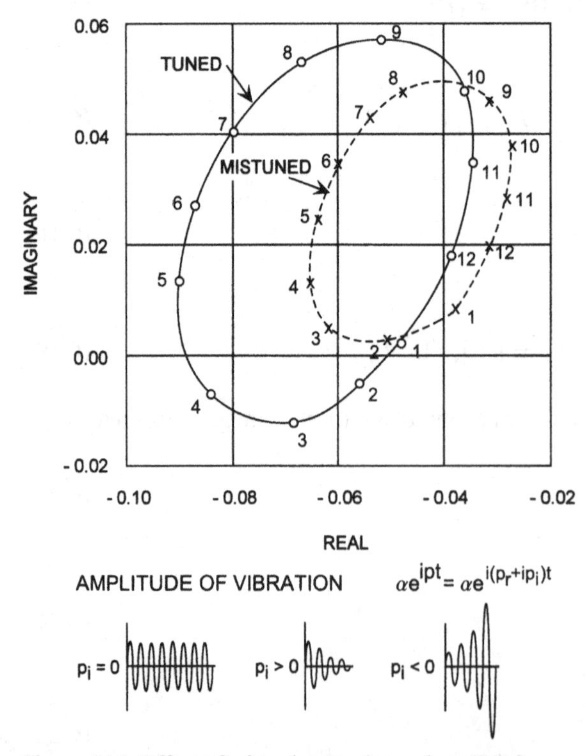

Figure 6.14. Effect of mistuning on eigenvalues (Srinivasan, 1997).

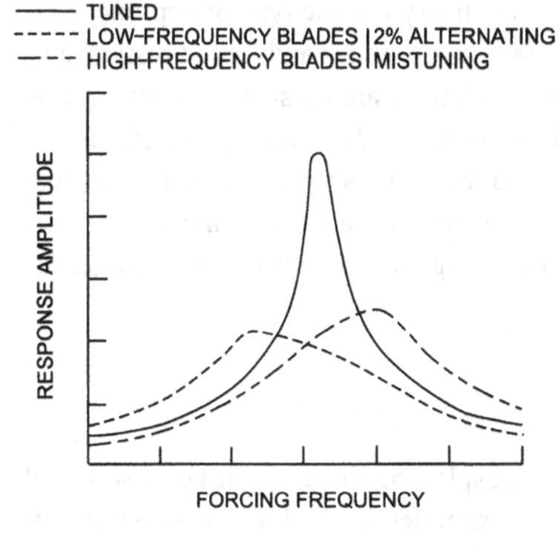

Figure 6.15. Forced response of a bladed-disk assembly to an eleventh engine-order excitation (after Kielb and Chiang, 1992).

which the susceptibility to aeroelastic instability can be determined. Imaginary parts are plotted against corresponding real parts for each interblade phase angle to obtain the locus. Note how the tuned system is vulnerable to flutter at modes corresponding to nodal diameters 2, 3, and 4. Upon mistuning, the locus shifts upward (toward more damping) and therefore the system has the potential to be aeroelastically stable, a clear case of exploiting the aeroelastic characteristics through mistuning.

Figures 6.15 and 6.16 display the response amplitudes that are possible for the same level and type of mistuning (alternate blades were mistuned such that the frequency was alternately high and low by 2% of the mean frequency). Note how different the responses can be if the forcing function changes in its distribution. The first set is a response for a forcing function with eleven harmonics around a rotor (eleventh engine order), whereas the second set is the same rotor responding to a forcing function corresponding to the fourteenth engine order. Thus, much more elaborate analyses are needed to determine the type of mistuning that may yield the desired result for an acceptable forced response. When an acceptable mistuning pattern is obtained, calculations need to be made to determine its susceptibility to flutter.

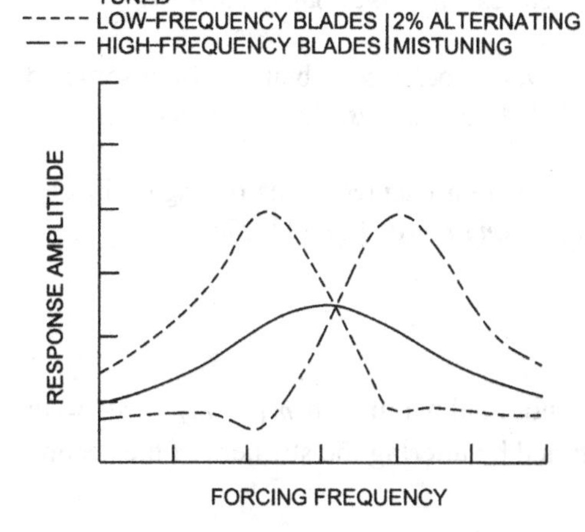

Figure 6.16. Forced response of a bladed-disk assembly to a fourteenth engine-order excitation (after Kielb and Chiang, 1992).

Researchers and designers in the jet engine industry use the type of approach discussed above to estimate the influence of mistuning on flutter and forced response of bladed-disk assemblies. The details of these calculations are outside the scope of this book, but interested readers may refer to Srinivasan (1997) to learn more about this topic and to reach a vast number of references to research papers. Suffice it to say that the analyses predict that mistuning blades provides a distinct advantage in regard to preventing the onset of flutter. However, blade mistuning may not always limit dangerous resonant stresses.

6.4 Summary

Mistuning structural systems by design aims at exploiting the dynamics of disordered systems to advantage. The jet engine related examples studied above show an uneven distribution of vibratory energy as evidenced by markedly varying amplitudes of vibration of blades around a turbine disk. Thus, some parts of the structure receive disproportionately larger energy than others, and therefore a design requirement must be to minimize the maximum amplitudes by careful selection of the extent and distribution of mistuning. Frequently, an optimization problem can be formulated and solved to determine the optimum amounts of mistuning to be introduced and their locations. When the mistuning of the substructures results from manufacturing variations, the problem may instead be to determine the best possible arrangement of a given population of parts; this often requires statistical analysis of the component properties and the system response. The potential to design and develop smart structures by deliberately mistuning components offers much promise as more advanced analytical tools are developed.

BIBLIOGRAPHY

Bergman, L. A., and D. M. McFarland. 1988. On the vibration of a point supported linear distributed system. *Journal of Vibration, Acoustics, Stress, and Reliability in Design* 110(4):485–492.

Bergman, L. A., and D. M. McFarland. 1989. Wide-band random excitation of beams and rectangular plates coupled to discrete substructures. *Structural Safety* 6:87–97.

Mignolet, M. P. 1996. Private communication.

Pierre, C., D. M. Tang, and E. H. Dowell. 1986. Localized vibrations of disordered multi-span beams. In *Proceedings of the 27th AIAA Structures, Structural Dynamics and Materials Conference*.

Srinivasan, A. V. 1997. Flutter and resonant vibration characteristics of engine blades. *ASME Journal of Engineering for Gas Turbines and Power* 119:742–775.

PROBLEMS

1. The 3-DOF system of Fig. 6.17 has the nominal parameters $m_1 = m_2 = m_3 = m$ and $k_1 = k_2 = k_3 = k_4 = k$. It is mistuned by altering the stiffness of the second

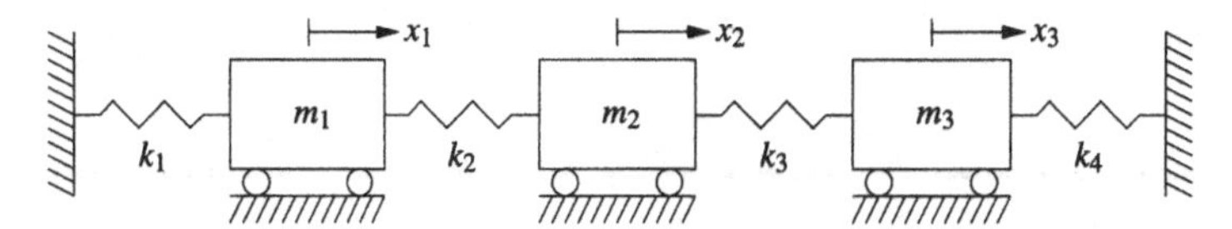

Figure 6.17. Nominally periodic 3-DOF system.

spring so that $k_2 = (1 + \varepsilon)k$. Calculate the effect of this mistuning on the participation of mass m_2 in the system's natural modes of vibration.

2. For the bladed-disk assembly shown in Fig. 6.13, assume the disk to be rigid. Further assume that each blade has a mass m, stiffness k, and a damping coefficient c. Let $f_j = Ae^{i(\omega t + j\beta)}$ represent the aerodynamic forces, where A is a complex number corresponding to $w_j = w_0 e^{i(\omega t + j\beta)}$, $\beta = 2\pi n/N$, n represents the harmonic vibratory pattern around the assembly, and N is the total number of blades. Under these assumptions derive the condition under which the flutter frequency is essentially identical to the blade's natural frequency. Also derive and discuss the conditions under which the system is on the verge of instability (flutter), completely stable, and completely unstable.

3. The solutions assumed above are in the form of traveling waves. An equivalent solution is to represent w_j in terms of two standing wave modes as follows: $w_j = q_c \cos j\beta + q_s \sin j\beta$. Show that the analyses based on this representation for the conditions stated in Problem 2 above lead to identical results.

4. How would you formulate the flutter problem using either standing or traveling wave representation if an event occurred in which blade number 1 is broken in half?

5. Eight masses are connected in a ring, each coupled to its immediate neighbors by springs and dashpots. Each element is nominally identical to all others of the same type, described by the three constants m, c, and k.

(a) Let $m = 0.25$ kg and $k = 3 \times 10^5$ N/m. Find c so that the system possesses 0.5% damping in its lowest mode.

(b) Let each mass be driven by a unit harmonic force of frequency ω. These forces act in phase, and ω varies over a range including the first three natural frequencies of the structure. Compute the peak displacements of each mass at each resonance.

(c) Choose several sets of random perturbations $\{\varepsilon_1, \ldots, \varepsilon_8\}$, $-0.1 \le \varepsilon_i \le 0.1$, $i = 1, \ldots, 8$. Setting the masses equal to the mistuned values $m_i = \varepsilon_i m$, repeat the analysis of part (b).

7

Fiber Optics

7.1 Introduction

An optical fiber "conducts" light in much the same way a copper wire conducts electricity. The most common use of fiber optics is in transmitting data such as telephone conversations, television programs, and numerical data over long distances. Data transmission takes place using light instead of electricity. In the context of smart structures the role of fiber optics is different and is in the form of sensing physical parameters such as strain, temperature, pressure, and vibration in structural components.

As the fibers can be embedded inside a structure, the deformation characteristics of the structure can be measured anywhere along the fiber. In the design of complex structural components, such as jet engine fan blades, much analysis and static testing are conducted to estimate "suitable points" to locate strain gages for use in spin tests. Nevertheless, the complexity of the modes can be such that these "suitable points" may, in fact, miss some critical strain locations. By embedding a grid of fibers, a more thorough strain-mapping can be obtained than is possible via discrete strain gages located on the surface of structures. According to Claus (1991), "displacements at temperatures up to 1000 °C have been measured using silica fibers and arrays of similar silica Fabry-Perot sensors have been used to map strains on the wing of an F–15 aircraft during simulated flight conditions." Unlike strain gages, embedded optical fibers can be used to monitor curing of composite structures, a feature most useful in assessing structural integrity of components. Fiber-optic sensors can also be configured to measure the internal chemical states in structures, such as the penetration of corrosion-causing de-icing salts in bridge decks.

7.2 The Physical Phenomena

7.2.1 Total Internal Reflection

The basic principle governing the operation of fiber optics in applications is *total internal reflection*, which can occur when light waves traveling in one medium encounter

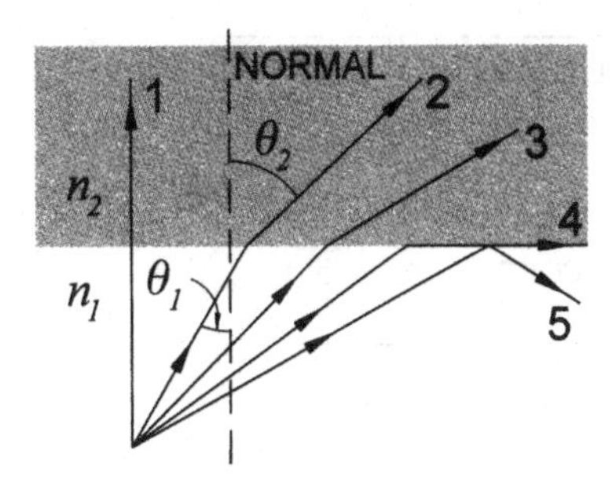

Figure 7.1. Total internal reflection (after Serway, 1992).

another medium. An example is light traveling in air encountering a surface of water. Because the refractive indices for water and air are different, the light waves behave differently at the interface. Consider medium 1 with a refractive index n_1. Light waves entering medium 1 encounter another medium of refractive index $n_2 < n_1$, as shown in Fig. 7.1. Note how a ray of light with an incidence angle θ_1 experiences refraction with an angle of refraction, θ_2. As the angle θ_1 continues to increase, the angle θ_2 also increases until $\theta_2 = 90\,^\circ$ for $\theta_1 = \theta_c$, where θ_c is defined as critical angle of incidence.

For $\theta_1 > \theta_c$, the ray of light is reflected back entirely into medium 1, which represents the condition for total internal reflection with no loss in transmission. The ray numbered 5 in the figure shows this condition and behaves as if it has struck a perfectly reflecting surface. Such rays obey the law of reflection so that the angles of incidence and reflection are equal. They also obey the law of refraction or Snell's law, which states that a ray undergoes refraction when it propagates from one medium to another such that the ratio of the sine of the incident angle and the sine of the refracted angle is equal to the ratio of the reciprocal of refractive indices of the media, that is, $\sin\theta_1/\sin\theta_2 = n_2/n_1$. Thus, the critical incidence angle θ_c may be calculated from $n_1 \sin\theta_c = n_2$ and, therefore, $\sin\theta_c = n_2/n_1$ for $n_2 < n_1$, which suggests that total internal reflection can occur only when light waves traveling in a medium with a certain index attempt to move into another medium with a lower refractive index.

As stated earlier, total internal reflection occurs when $\theta_1 > \theta_c$. Clearly, rays of light propagating from a dense medium through successive layers of less dense media will continue to bend and experience total internal reflection. The principle of total internal reflection can be used to "pipe" light from one location to another. In a glass rod or fiber, light entering the glass medium at one location is trapped as it were, as shown in Fig. 7.2, and with successive reflections continues to travel through the fiber entirely confined.

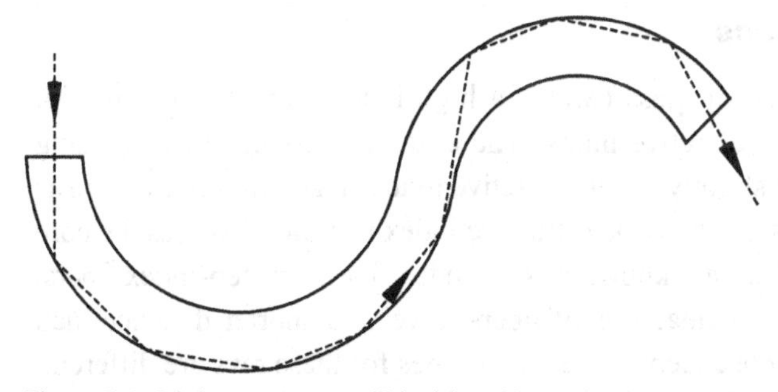

Figure 7.2. Light rays "trapped" inside a fiber (after Urone, 1998).

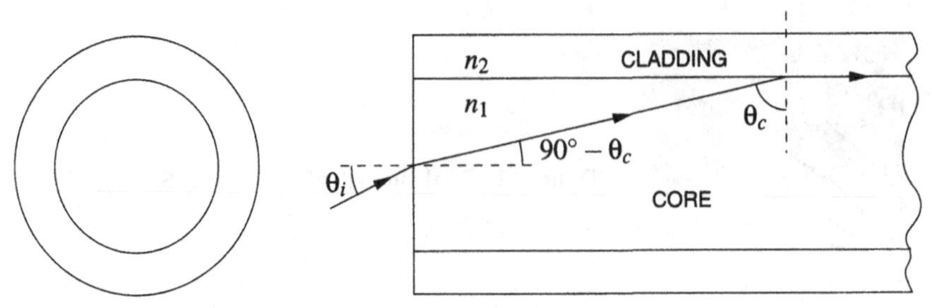

Figure 7.3. Numerical aperture (after Kao, 1998).

This physical phenomenon serves as the foundation for a wide variety of engineering uses of fiber optics. An optical fiber can be viewed as a light guide obeying Snell's law. The fiber has a core through which light waves can propagate by total internal reflection from a cladding of lower refractive index.

7.2.2 Numerical Aperture

A relationship between the incident angle, θ_i, and the indices of refraction may be derived for the condition of total internal reflection as shown below with reference to Fig. 7.3.

$$\frac{\sin\theta_i}{\sin(90° - \theta_c)} = \frac{\sin\theta_i}{\cos\theta_c} = \frac{n_1}{n_2}. \tag{7.1}$$

Also,

$$\sin\theta_c = \frac{n_2}{n_1}. \tag{7.2}$$

Solving (7.1) and (7.2) for $\sin\theta_i$ gives

$$\sin\theta_i = \left[\left(\frac{n_1}{n_2}\right)^2 - 1\right]^{1/2} \tag{7.3}$$

The quantity $\sqrt{(n_1/n_2)^2 - 1}$ is defined as numerical aperture, which establishes the maximum angle of incidence that assures total internal reflection in the fiber.

7.3 Fiber Characteristics

Optical fibers are usually made of glass (SiO_2), a high-index material, mixed with various dopants to control the refractive index. The core is surrounded by cladding material, which is also glass of slightly lower refractive index. The difference can be as small as 0.001 or 0.002. Fibers in which the refractive index is uniform across the core thickness, as shown in Fig. 7.4(a), are known as step-index fibers. In step-index fibers, rays traveling close to the longitudinal axis of fibers traverse a shorter distance than those at an angle to the axis. Consequently, the travel times for these rays are different,

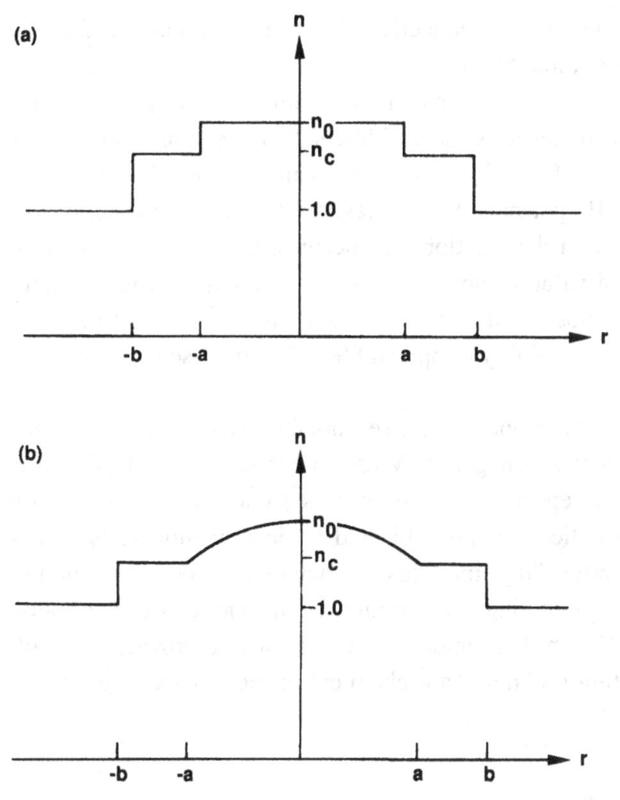

Figure 7.4. Index profiles for step-index and graded-index fibers. (Courtesy Peter Cheo, 1985.)

leading to light dispersion. However, if the core is made of glass with refractive indices varying smoothly across the core diameter as shown in Fig. 7.4(b), then the travel distances can be controlled in such a way that the light dispersion is minimized.

Although the discussion above introduces the concepts of light transmission through optical fibers, it is somewhat incomplete because it does not include oscillations along the other principal directions. In fact, propagation of light along the optic fibers can be looked upon as electromagnetic waves with a wide variety of modes. Thus it should be noted that even with an optimum grading of refractive index through the thickness, fibers that are truly multimode exhibit inevitable dispersion. However, with drastic reduction in core thickness, one can reduce the number of modes that can be physically supported by the fiber. For example, a core with a 10 μm diameter is only about four wavelengths across and therefore cannot allow any more than a single mode, thus avoiding the dispersion problem.

The cores of optical fibers are made of extremely low-impurity silica and the cladding is typically fluorine-doped glass 125 μm in diameter. In view of the lower specific gravity of silica compared to that of copper (2.2 vs. 8.9), the diameter of an optical fiber can be less than a tenth of the diameter of a corresponding copper wire to convey the same information. The consequent weight reduction, for example, in a large aircraft communication system can be a saving of 1 000 lb (see Lacy, 1992). In addition, fiber optics possess an enormous capacity to carry information because of a vast bandwidth, which is an incredible 50 THz.

Typical single mode fibers have core diameters of about 5 μm, which compares favorably with the 1 to 3 μm diameter for structural fibers. However, multimode fibers are much larger in size and may be 100 μm to 200 μm in diameter. Ultimate tensile strengths of the order of 10^6 psi have been measured for short test sections of freshly drawn glass fibers. A slight degradation may occur in the manufacturing process leading to a compromised value of about of 800 000 psi. With a Young's modulus of 10×10^6 psi, which is close to that of aluminum, and the ability to survive strains of the order of 8%, the applicability of optical fibers for strain-sensing becomes evident.

With appropriate adhesives, fiber-optic strain sensors can have long life – much longer than that of conventional strain gages. With proper selection of adhesives and coatings, optical fibers are reported to have sustained one million cycles in a simulated graphite-reinforced helicopter tail subjected to delamination tests. Polyimides that have a high temperature/high modulus characteristic have been found to be ideal in providing a thin, hard coating and the increase in fiber diameter because of the coating can be a mere 10 μm. Polyimides are also known to provide good adhesion to epoxies, demonstrating no significant chemical or mechanical degradation (Dunphy, Meltz, and Morey, 1995).

7.4 Fiber-Optic Strain Sensors

7.4.1 Strain Measurement

The basic design philosophy of engineering structures has changed drastically from a conservative "safe life" approach to a more scientific "damage tolerant" approach. "Safe life" was set on the basis of fatigue test data, the extent of whose scatter determined setting the safe limit below the lowest values. With this approach, components that may have many additional years of useful service had to be retired. Advances in the science and technology of fracture mechanics provided a more rational basis to view structural failure on the basis of size and propagation characteristics of cracks. Thus, it was argued that the mere presence of cracks in a structure could not be the basis to declare that the product has outlived its usefulness; rather, crack lengths and propagation characteristics must determine component life. The success of the latter approach is dependent on the accuracy with which structural components with defects are modeled and tested in order to generate a basis in regard to critical crack lengths. It therefore follows that the ability to monitor cracks and their dynamics will be mandatory to ensure safe operation of structural systems. Embedded fiber optic sensors are invaluable in such monitoring and diagnostic tasks.

In a broad classification, sensors are described either as extrinsic or intrinsic. In extrinsic fiber sensors, light senses the parameter under study (strain, temperature, pressure, vibration, chemical concentration, or other phenomena) at some location along the fiber and exits the sensor either by reentering the input fiber or by entering another fiber to reach a detection device. In intrinsic sensors, on the other hand, changes in one or more optical parameters (intensity, phase, polarization, wavelength, frequency, timing, or modal content) are observed as light, propagating through the fiber, experiences change due to the influence of the measurand.

Only the basic principles that govern the unique ideas implicit in the arrangements of several sensors are described below. Of these, the most promising are perhaps the Bragg grating and the white light interferometer (Huston, 1999). Additional details on these and other sensing mechanisms are available in the list of references included here.

7.4.2 Microbent and Graded-Index Fibers

In a structure embedded with fiber optic sensors, the deformation of the structure imparts strain to the elastic fiber. As a result, changes occur in certain optical characteristics, such as optical intensity and phase. A measurement of phase change is the common approach and provides a measure of deformation. In a sense, one could visualize changes in the length of the fiber resulting from deformation, which in turn changes the distance of "flow" of light. This change in length could further be visualized as the basis of arrival time of the light wave form that is contained in the phase information.

In one such measurement mechanism, phase changes are measured between a reference signal through a fiber and another signal coming through a similar fiber that contains microbends along its length (see Fig. 7.5). These microbends, which are intentionally built along the fiber at periodic intervals, allow light to radiate out. The behavior of the light signal through the fiber with microbends is influenced by temperature, acceleration and strain. Thus, a comparison between a reference signal and a signal through its corresponding fiber with microbends provides a measure of the parameter of interest. In another measurement mechanism, sensing strain using graded index (GRIN) multimode fibers reduces the problem from phase detection to intensity detection (Giallorenzi, 1982), as these types of fibers exhibit substantial microbending losses. This is desirable because intensity detection is easier than phase detection.

A GRIN fiber that is periodically microbent attenuates light transmitted through it. The attenuation in GRIN fibers is more pronounced if the periodic spacing is

$$\Gamma = \frac{2\pi a}{\sqrt{2\Delta}}, \tag{7.4}$$

Figure 7.5. A schematic illustrating microbending in an optical fiber (Mutalik et al., 1994).

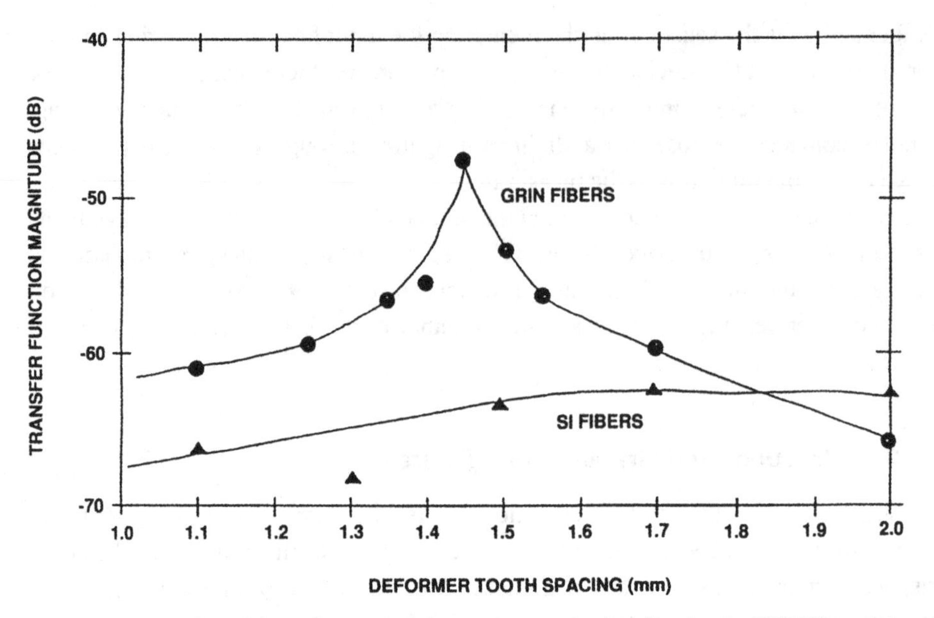

Figure 7.6. Illustration of the optimum pitch for maximum microbending for the GRIN fibers (Mutalik et al., 1994).

where a = radius of the core and Δ = difference in refractive indices of the core and the cladding. GRIN fibers embedded in composites undergo microbending when transverse stress is applied and therefore detect applied stress on the composite (Measures, 1989).

GRIN fibers embedded in composites derive microbending and therefore their sensitivity from the spacing between the structural fibers. GRIN fibers are expensive compared to step index (SI) fibers. However, SI fibers are less sensitive to applied stress compared to GRIN fibers, as shown in Fig. 7.6. One must note these advantages and disadvantages in the choice of fibers for use in individual applications.

7.4.3 Extrinsic Fabry-Perot Sensors

The basic principle that governs the operation of Fabry-Perot sensors can be understood with reference to Fig. 7.7. The instrument is constructed by providing an air gap (typically 4 mm in length and measured to an accuracy of $\pm 5\ \mu$m) between a single mode fiber and a reflection from a multiple mode fiber. The sensor is attached to a structural component through adhesives that faithfully transmit any deformation to the sensor leading to a change in the length of the gap. This, in turn, causes a phase change between the light of the reference signal (reflected at the glass-air interface at the left end of the gap in the figure) and the light from the sensing signal (reflected at the air-glass interface at the right end of the gap) because of interference between the two reflections. This phase change is a measure of motion at the gap location and serves as the basis for an accurate measurement of strain.

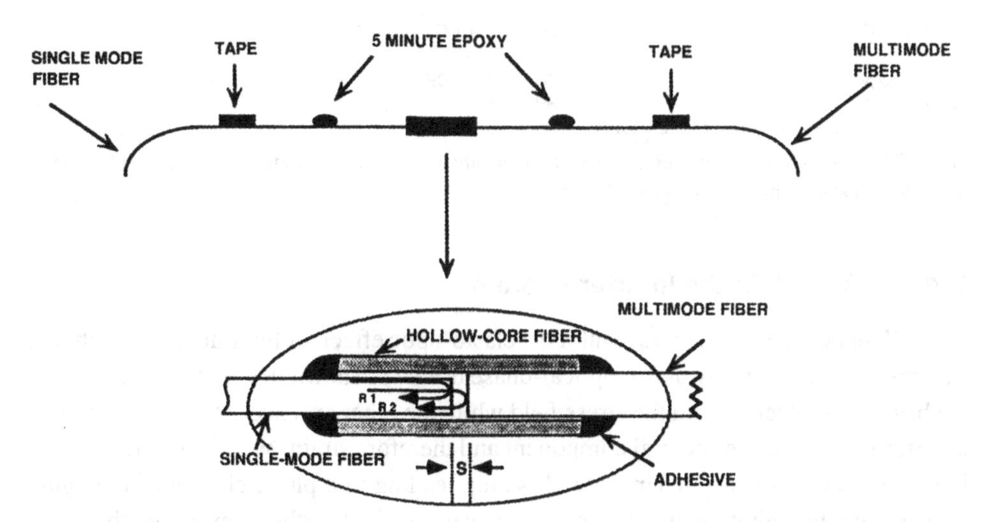

Figure 7.7. Fabry-Perot sensors. (Reprinted, by permission, from Experimental study of embedded fiber-optic strain gauges in concrete structures, Masri et al., 1994. Copyright American Society of Civil Engineers.)

An excellent example of the application of this instrument to measure strains in an F–15 airframe during a 10 g load fatigue test is described by Murphy et al. (1991). The results are depicted in Fig. 7.8. Fabry-Perot sensors demonstrated minimum detectable motion of the order of 0.1 nm and a strain of 0.01 μstrain. This is remarkable as it represents a sensitivity improvement of more than two orders of magnitude compared to foil gages. The author reports negligible hysteresis, and the difference between the measurements made by the fiber sensor and the strain gages was found to be less than 1% over the range of the tests.

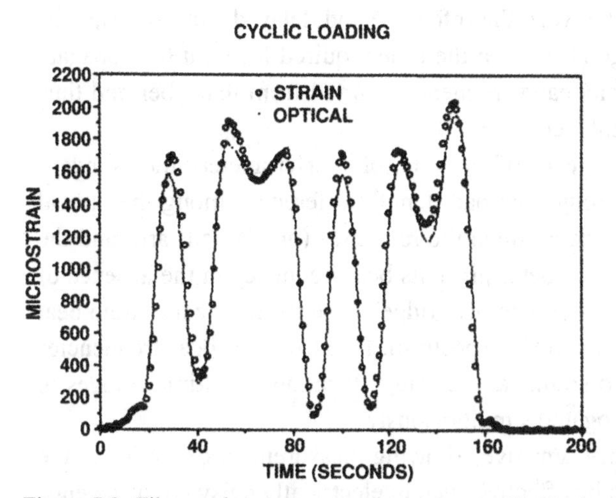

Figure 7.8. Fiber-optic sensor measurement on an F–15 aircraft. (Reprinted, by permission, from Experimental study of embedded fiber-optic strain gauges in concrete structures, Masri et al., 1994. Copyright American Society of Civil Engineers.)

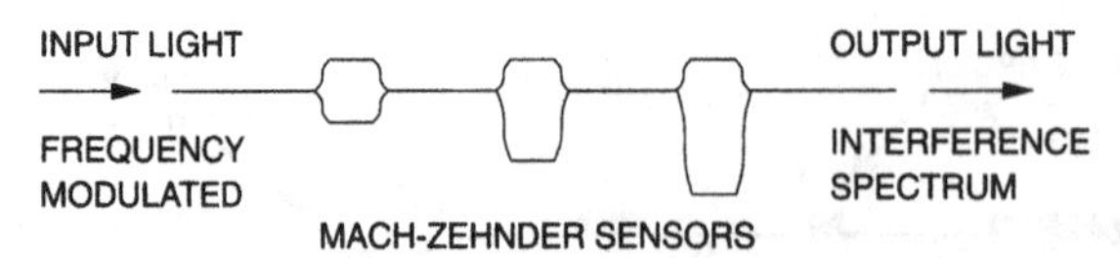

Figure 7.9. Mach-Zehnder interferometers. Distinct path differentials allow the sensors to be identified in the output spectrum (after Rogers 1987, p. 148).

7.4.4 Mach-Zehnder Interferometers

Mach-Zehnder interferometers exhibit the elasto-optic effect (optics influenced by fiber deformation) through a two-arm optical-phase bridge as shown in Fig. 7.9. One arm of the bridge is protected from the stress field while the other arm is exposed to the effect of deformation of the structural component and therefore strains with it. The refractive index of the exposed arm changes with strain, leading to a phase change of the light propagating through it relative to the light transmitted in the other arm. When the light waves are recombined, the combined wave amplitude will be influenced by the phase difference and serves as a measure of the strain amplitude.

Measurements can be made at discrete points along the body of the structure by a series of two-arm bridges. Both fibers are illuminated by a continuous-wave laser. The laser is amplitude modulated by another signal, so that the light propagating through the fibers has components at the laser frequency and at sideband frequencies determined by the frequency of the modulating signal. Furthermore, the frequency of the modulating signal is swept in time, typically as a ramp or sawtooth function, and the frequencies of the sideband signals vary at this sweep rate. Any difference in path length (more properly, propagation time) between the two fibers will lead to a difference in the sideband frequencies of the light emerging at the detector end of the fibers. The detector is a nonlinear circuit, such as a rectifier, and its output will include frequencies equal to the difference or beat frequency between the reference and delayed sideband signals. When the sensing arm of a bridge is strained the time required for light to propagate through it changes, altering the sideband frequency emitted from that fiber and thus causing a shift in the observed beat frequency.

Varying the length of one arm in each of the series of interferometers allows indentification of individual sensors through the path-length difference among the optical bridges. A fixed difference in length between the reference (unstrained) arm and the sensing (strained) arm of each bridge determines its beat frequency in the absence of strain, and by varying the length differences of bridges in series their zero-strain beat frequencies may be separated in the output spectrum. Because their beat frequencies vary independently in response to strains at the bridge locations, multiple bridges in series can be identified and interrogated simultaneously.

Mach-Zehnder sensors are very sensitive, allowing measurement of strains of the order of 10^{-8}, and are reported to be effective even in electrically noisy environments. As gages are added in series, however, the output spectrum can become hard to interpret because of the appearance of harmonics of the ramp sweep function.

7.4.5 Other Fiber-Optic Strain Measurement Techniques

7.4.5.1 Bragg Grating Sensor

Bragg gratings are extremely close parallel lines "written" onto a small length (1 to 20 mm) of the core of a fiber so as to create a systematic perturbation of the core's refractive index (Dunphy et al., 1995). This spatially periodic variation of the index of refraction acts as a filter by reflecting certain wavelengths while allowing others to pass through. Typical gratings have resonant wavelengths of 400 to 2 000 nm.

When broadband light traveling in the optical fiber encounters the grating, light at a wavelength proportional to the Bragg spacing is reflected back. This may be observed as a gap in the spectrum of the transmitted light, or as a peak in that of the reflected light. When the grating is strained, the Bragg spacing and thus the reflected wavelength change accordingly; the resulting change in the spectrum is used as a measure of strain. This is a robust sensing scheme because light wavelength does not change as it passes through other fibers or connectors.

7.4.5.2 White Light Interferometry

In this technique, a strained (active) fiber and an unstrained (reference) fiber are spliced together and broadband (white) light is sent into both. The light reflected from the ends of the two fibers will interfere in a manner that depends on their relative lengths, and so varies as the active fiber is strained. The change in the interference pattern is measured as an indicator of strain.

These fibers have proven to be stable over time under static loads. They are used in civil and geotechnical engineering applications that require strain measurements with long-term stability (Huston, 1999).

7.5 Twisted and Braided Fiber-Optic Sensors

As mentioned above, cladded glass fibers lose optical power from their core under microbending; lost optical power and microbending are quantitatively related. When optical fibers are twisted and braided (Fig. 7.10), the microbending and the consequent optical loss resulting from a small amount of ordinary bending of the fiber are magnified. Because the microbending is inherent, such braided optical fibers can be embedded in any pattern inside a composite, and therefore the performance is independent of the spacing of structural fibers. The process of braiding can be accomplished as shown in Fig. 7.11. First, one end of the fiber is fixed and the other end of the fiber is twisted. After an appropriate amount of twisting, the fiber doubles on itself and is allowed to

Figure 7.10. Schematic of twisted and braided fiber structure (Mutalik et al., 1994).

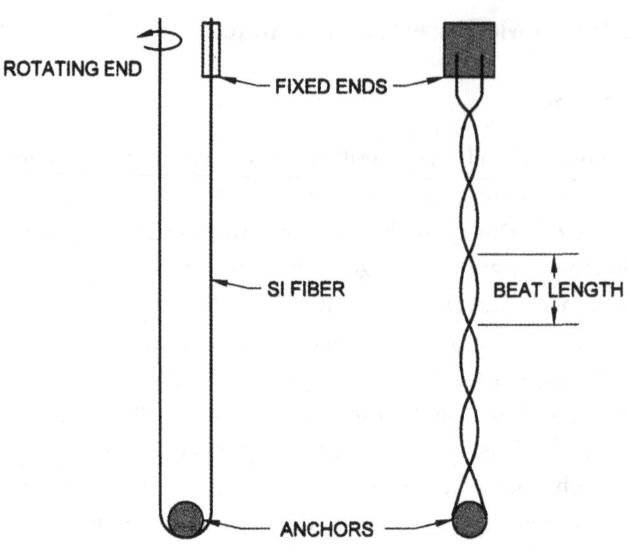

Figure 7.11. The process of making twisted and braided fiber structure (Mutalik et al., 1994).

braid automatically. For a given length and type of fiber, the number of rotational twists determines the braid pitch.

7.5.1 Preliminary Experiments

In tests conducted on individual strands (EB100/125/140/250–0.14NA, acrylic clad, step index multimode), the fiber was found to have a Young's modulus E of 2.5×10^3 ksi and an ultimate strength S_u of 81.9 ksi (Mutalik et al., 1994). The latter is much less than that of Kevlar–29 (1700 ksi).

Optical power to the fibers was supplied during uniaxial tests by a Hewlett-Packard HFBR–1604 source operating at 670 nm (visible red). The source had a connector receptacle and the input end of the fibers was fitted with an FC connector and the EBOC TK–6 HCS® fiber Termination Kit and coupled to the laser source. A careful note was made of the optical transmission as a function of applied load using the Model 835 SL Newport Optical Power Meter. The load was then converted by standard techniques into microstrain and the optical transmission plotted against the microstrain as shown in Fig. 7.12. The results were repeatable and confirmed the potential of braided fibers as sensitive strain sensors in composites.

7.5.2 Coupon Tests

A coupon with optical fibers embedded in epoxy was developed to verify the postulate that twisted and braided fibers could act as efficient strain sensors. The "dog bone" configuration coupon shown in Fig. 7.13 was 0.25 in. thick, 8 in. long, and 1.5 in. wide for use in a three-point bending test. Only the middle 5-in. section of the epoxy was subjected to bending. A photograph of this arrangement is shown in Fig. 7.14. Stress

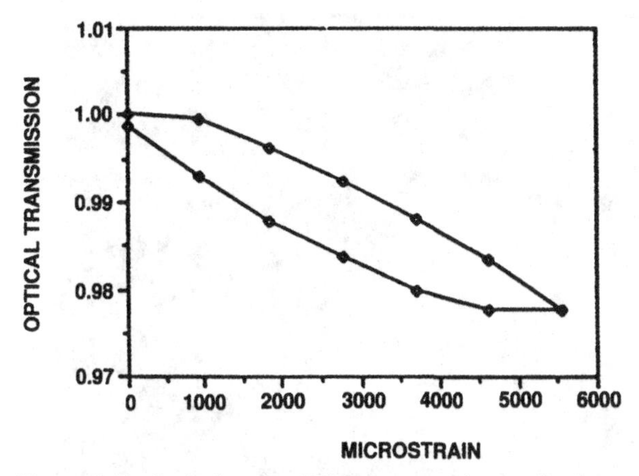

Figure 7.12. Optical transmission versus microstrain characteristics of SI fibers in uniaxial tests (Mutalik et al., 1994).

developed due to bending in turn induced microbending losses in the fibers, and these were measured. The load changes were recorded from the digital readout of the testing machine in real time and the maximum load on the sample was allowed to be about 150 lb. The load was then converted into microstrain by using simple beam theory and by using the value of the Young's modulus of the epoxy, which had been determined independently.

The tests performed on the coupon generated hysteresis patterns similar to those observed with the individual braided fibers. However, polymide-coated fibers, tested in the same manner as the earlier fibers, resulted in very little hysteresis and a more linear range of optical transmission with respect to applied strain (Fig. 7.15). In addition to a more linear regime free of hysteresis, the polymide-coated fibers have several other distinct advantages. Their Young's modulus is 9.25×10^3 ksi and their ultimate strength 383 ksi. This is much higher than the corresponding values for the acrylic coated fibers used in the coupon. Further, these values compare favorably with materials like E-Glass ($E = 10.5 \times 10^3$ ksi, $S_u = 500$ ksi) and Kevlar–29 ($E = 9.4 \times 10^3$ ksi, $S_u = 1700$ ksi). The polymide-coated fibers can also handle the higher temperatures frequently encountered during the composite curing stages.

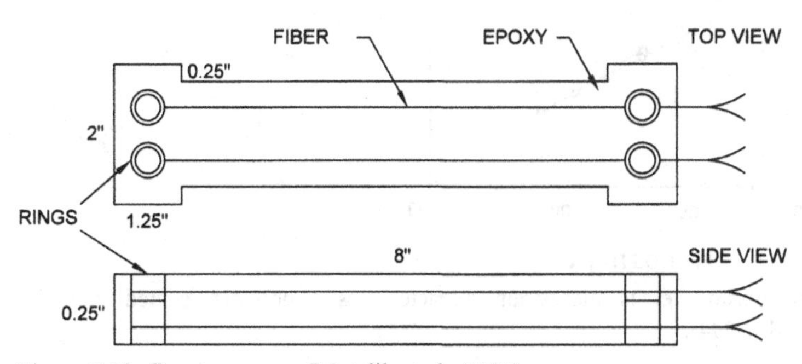

Figure 7.13. Epoxy coupon (Mutalik et al., 1994).

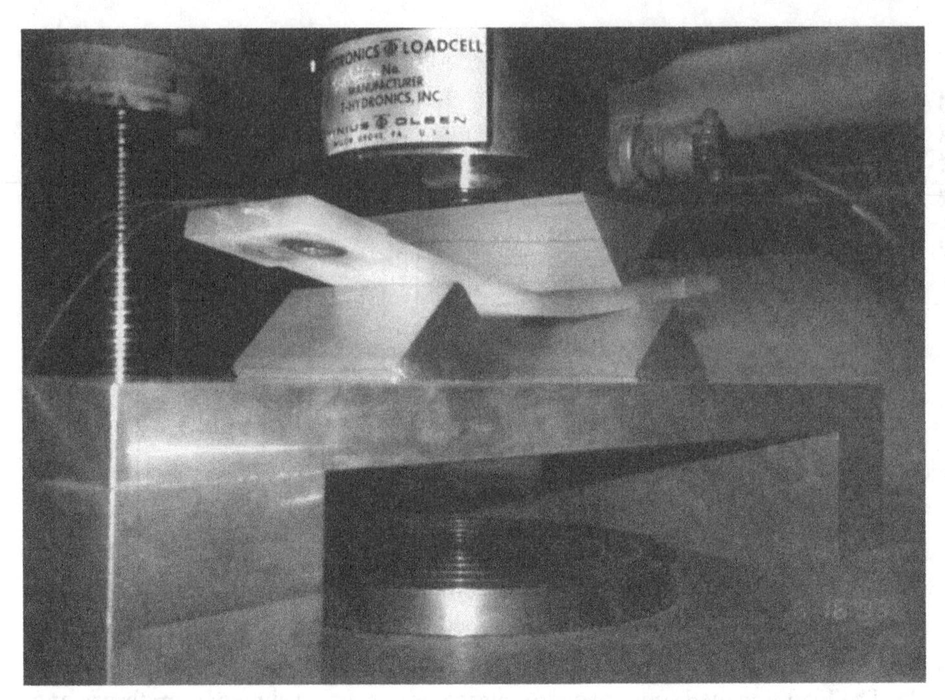

Figure 7.14. A photograph of the 3-point bending test in progress (Mutalik et al., 1994).

7.6 Optical Fibers as Load Bearing Elements

Embedded fibers are useful as sensors merely because they experience deformation because of the bond with the structure. Thus, optical fibers are in fact carrying the loads acting upon the structure. In the experiment reported earlier (see Fig. 7.14), the plate is clearly experiencing large displacement and still the fibers appear to have retained

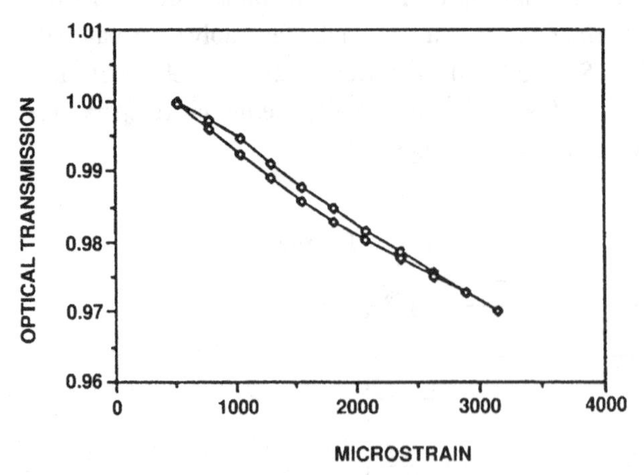

Figure 7.15. Optical transmission versus microstrain characteristics of polymide-coated fibers in uniaxial tests (Mutalik et al., 1994).

their integrity. Therefore, a question arises in regard to possible multifunctional characteristics of optical fibers: why can't they function *both* as sensors *and* load carrying members? This question can be answered only through a research program, but it is believed that a key element contributing to the success of this concept is the diameter of the optical fibers, which is determined largely by the type and diameter of cladding and coating.

7.7 Additional Applications

7.7.1 Crack Detection

When a fiber-optic strain sensor experiences strain along with the structure to which it is attached, neither sustains damage in the course of normal operation. On the other hand, an optical fiber may be used to detect localized damage to its supporting structure, for example by embedding it in a composite in such a way that it fails along with the surrounding matrix material. One possible configuration is shown schematically in Fig. 7.16, where a crack in a structural member has led to a crack in the embedded fiber at the same location.

The development of cracks can be detected by monitoring the intensity of light transmitted through the embedded fiber. Light will be lost from a crack in the fiber, and this loss will increase as the crack opens. This reduction in transmitted light will reveal the presence of a crack, but not its location within the structure.

Both the existence and the location of a crack can be determined by utilizing a more sophisticated interrogation method. In optical time domain reflectometry (OTDR), a very brief pulse of light is sent into the embedded fiber, where it propagates normally until it reaches the crack. Part of the incident pulse is reflected from the crack and travels back toward the input end of the fiber. When such a reflected pulse is detected, a crack is known to exist. Furthermore, because the speed of light in the fiber is known, the time between the generation of the input pulse and the arrival of the reflection can be used to compute the location of the crack.

7.7.2 Integration of Fiber-Optic Sensors and Shape Memory Elements

The sensitivity of optical fibers to sense strains in a structure and the ability of shape memory alloy wires to actuate offer an unusual opportunity to combine the two in

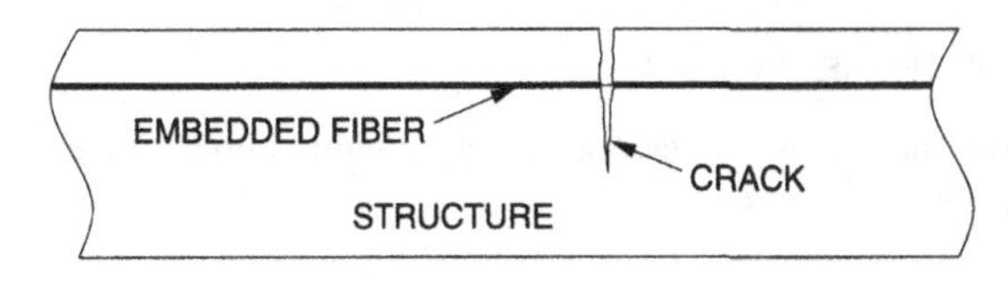

Figure 7.16. An embedded optical fiber will be damaged along with the surrounding material.

structural applications. A grid pattern of optical fibers can be visualized from which not only the strain level but also the location of undesirable strain may be established. With this information, actuation of the structure can be initiated by energizing the corresponding shape memory alloy wires grid resulting in a smart structure.

Aircraft wings, bridge decks, and buildings are some of the examples in which a judicious combination of these two technologies may contribute to a greater level of structural integrity than otherwise possible. Although this approach is conceptually attractive, the technology needs to be developed through research efforts. The issues pertain to the use of both fibers (shape memory, optical) in addition to structural fibers in a composite structure. Structural weight and stress concentration are to be balanced against cost questions. Feasibility studies are underway, and if successful, the impact on the design of smart structures is evident. An extremely revolutionary development may include the potential use of optical fibers as sensors and load-bearing elements. Clearly, a success in such an attempt will have enormous implications on the design and development of structures in the future.

7.7.3 Chemical Sensing

Structural materials, including metals, concrete, and some composites, can fail following chemical changes or chemical attack. Embedded fiber-optic sensors may be used to measure the concentrations of various chemical species within a structure (Huston, 1999). Most such sensing techniques, such as the arrangement shown in Fig. 7.17, are spectroscopic. The sensor contains a chemical that changes color upon reacting with that penetrating the structure. This color change is detected as a shift in the spectrum of light transmitted through the sensor.

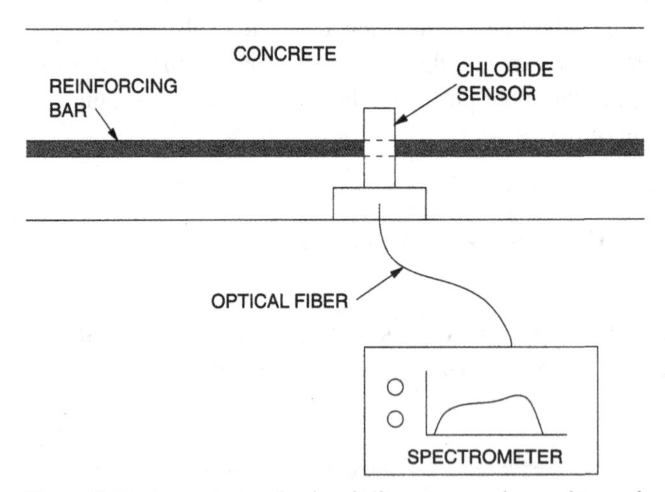

Figure 7.17. A spectroscopic chemical sensor may be used to evaluate the penetration into a bridge deck of damaging chlorides from de-icing salts.

7.8 Summary

Earlier we stated that a structural component is deemed "smart" if it is capable of sensing a change and adapt itself appropriately in order to preserve its structural integrity. Therefore, a smart structure must be able to monitor itself throughout its lifetime and obtain an accurate estimate of its integrity. This necessarily defines the need for a distributed sensing medium. Conventional strain or temperature sensors are localized and not very suitable in this context. Occasional testing of structures, such as ultrasonic fatigue tests, are expensive and do not give real time defect information. Optical fibers appear to offer the preferred means of sensing changes in smart structures as and when desired. In addition, the potential of these fibers to function as load carrying elements in a structure may lead to vastly superior, reliable, and cost-effective smart structures.

BIBLIOGRAPHY

Bertholds, A., and R. Dändliker. 1987. Deformation of single-mode optical fibers under static longitudinal stress. *Journal of Lightwave Technology* LT-5(7):895–900.

Claus, R., ed. 1991. *Proceedings of the Conference on Optical Fiber Sensor-Based Smart Materials and Structures*, Virginia Polytechnic Institute.

Dunphy, J. R., G. Meltz, and W. W. Morey. 1995. Fiber optic smart structures. In *Optical Fiber Bragg Grating Sensors: A Candidate for Smart Structure Applications*. New York: John Wiley & Sons, Inc.

Giallorenzi, T. G. 1982. Optical fiber sensor technology. *IEEE Journal of Quantum Electronics* QE-18(4):457–458.

Huston, D. 1999. Private communication.

Kersey, A. D. and A. Dandridge. 1990. Applications of fiber sensors. *IEEE Transactions on Components, Hybrids and Manufacturing Technology* 13(1):137–143.

Lacy, E. A. 1992. The ultimate in tele-communications. In *Physics for Scientists & Engineers* 1009. Philadelphia: Saunders College Publishing.

Lerner, E. J. 1997. Optical fibers carry information of the age. *Laser Focus World* 33(3): 101–109.

Masri, S. F., M. S. Agbabian, A. M. Abdel-Ghaffar, M. Higazy, R. O. Claus, and M. J. de Vries. 1994. Experimental study of embedded fiber-optic strain gauges in concrete structures. *ASCE Journal of Engineering Mechanics* 120(8):1696–1717.

Measures, R. M. 1989. Fiber-optics smart structure program at UTIAS. In *Proceedings of the SPIE: Fiber Optic Smart Structures and Skins II*, 1170.

Murphy, K., M. Gunther, A. Vengsarkar, and R. Claus. 1991. Fabry-Perot fiber optic sensors in full scale testing on an F-15 aircraft. In *Proceedings of the SPIE O/E Fibers Conference*, Boston: SPIE.

Mutalik, V. G., A. V. Srinivasan, H. Canistraro, M. Hodge, and C. Roychouduri. 1996. Twisted and braided fiber-optic sensors in smart structures. In *Proceedings of the Second International Conference on Intelligent Materials and Structures*, Williamsburg, VA.

Pascual, J. 1991. Uniaxial mechanical performance of smart composite structures with embedded optical fibers. Master's thesis, The Pennsylvania State University.

Pigeon, F., et al. 1992. Optical fiber Young modulus measurement using an optical method. *Electronics Letters* 28(11):1034–1035.

Serway, R. A. 1992. *Physics for Scientists & Engineers*. Philadelphia: Saunders College Publishing.
Smith, H. Jr. et al. 1989. Smart structures concept study. In *Proceedings of the SPIE: Fiber Optic Smart Structures and Skins II* 1170:224–229.
Turner, R. D., T. Valis, W. D. Hogg, and R. W. Measures. 1990. Fiber-optic strain sensors for smart structures. *Journal of Intelligent Material Systems and Structures* 1(1):26–49.
Urone, P. P. 1998. *College Physics*. Belmont, CA: Brooks/Cole Publishing Company.

PROBLEMS

1. A Fabry-Perot sensor with a gap of 4 mm is used to measure a strain of 100 μstrain in the structure to which it is bonded. Find the time and phase differences between the reference and sensing reflected signals.

2. A Mach-Zehnder strain bridge operates with a ramp sweep signal of 10 GHz varying the wavelength of the input laser light from 500 to 600 nm. The active arm of the interferometer is exposed to the 100 μstrain deformation of Problem 1 over a gage length of 10 mm.

(a) Find the wavelength of the resulting line in the interferometer's output spectrum.

(b) Calculate the sensitivity of this bridge.

(c) How is sensitivity related to gage length in this configuration?

3. Two Mach-Zehnder sensors of the type described in Problem 2 are to be connected in series.

(a) What path length differential should be introduced in order that their outputs can be distinguished?

(b) What will be the resulting shift in wavelength?

4. An embedded fiber-optic sensor is used to detect cracking in a large composite structure via time-domain reflectometry. If independent reflected signals arrive at the input end of the fiber at times $t_1 = 300$ ns, $t_2 = 450$ ns and $t_3 = 1250$ ns after injection of the interrogating pulse, find the number of cracks and their distance from the input end of the fiber.

8

Control of Structures

8.1 Introduction

In categorizing smart structures as open-loop or closed-loop, we have implied that some are capable of responding actively to changes in their state or environment, for example by altering their effective mechanical properties. Much of this book is devoted to analyzing means for sensing or inducing mechanical phenomena, such as force or displacement, often through the use of electrical signals. In this chapter we shall consider the devices or subsystems that generate the signals to which the actuators respond and the effects they produce in some common configurations.

In a closed-loop structure, sensor outputs are processed by the controller to generate actuator commands. Open-loop structures may employ neither sensors nor actuators; on the other hand, nothing prevents the use of actuators without feedback (see, e.g., the multiplexed SMA wire actuator experiment reported in Chapter 3). Such an open-loop smart structure still needs a controller to generate the signals applied to its actuators. Thus, a closed-loop smart structure *requires* a controller, while an open-loop smart structure may or may not incorporate one. We will be concerned here primarily with feedback systems, that is, closed-loop structures.

It will come as no surprise that the introduction of feedback can radically alter the dynamics of a structural system, affecting its natural frequencies and modes, its transient response, and even its stability. Fortunately, in studying smart structures we can take advantage of the literature in the field of structural control, which has matured greatly over the last two decades. Much of this work has dealt with conventional structures modified by the addition of discrete sensors or actuators, as opposed to the more fully integrated systems implied by the term smart structure, but it has gone far toward connecting the sometimes disparate fields of structural dynamics and control theory.

8.2 Structures as Controlled Plants

We consider in this section some points that must be addressed in applying automatic control techniques to alter the dynamic response of a structure. Our focus is on those features of structural dynamic models that distinguish them from more typical plants and thus potentially complicate the controls problem.

8.2.1 Modeling Structures for Control

In discussing control theory in the abstract, it is common to assume the availability of a model of the plant to be controlled, typically in the form of a differential equation or system of differential equations (i.e., an input-output relation or state-space model). For many purposes, the number of inputs and outputs or the number of states can be regarded as a parameter that is defined on a case-by-case basis, and theoretical developments can proceed under the assumption that this number is "small" in the sense that the sheer size of the matrices to be manipulated will present no numerical difficulties, for example. This approach may work well for mechanical systems with a few degrees of freedom, even bearing in mind that each degree of freedom gives rise to two state variables (one displacement and one velocity).

However, it is common to perform static and dynamic analyses on structures with many degrees of freedom. When a structural dynamic model results from the finite element discretization of a large or complex structure, it may have tens of thousands of degrees of freedom. Although numerical methods to compute the natural frequencies and natural modes of such a system are available, the computational tools of modern control cannot all be expected to be so robust. (This is due to the good numerical properties of most structures' coefficient matrices as well as to the types of problems historically of greatest interest to the controls and structures communities.) A consequence of this disparity is that the straightforward reduction of even a moderate structural model to state-variable form can produce a state-space model of ungainly dimensions. This has motivated interest in reducing the size of structural models while retaining acceptable accuracy, and in exploiting the structure of the equations governing mechanical systems.

The idea behind *model reduction* or *reduced order modeling* is to identify those degrees of freedom or state variables least relevant to the goals of analysis and systematically eliminate the corresponding equations from the model. This may mean isolating those degrees of freedom that contribute least to the forced response or those states that have little effect on the control law. In some cases it is possible to determine by inspection which equations may best be dispensed with, for example, those corresponding to in-plane motion when out-of-plane bending is the deformation mode of interest. More generally, though, a mathematical criterion must be devised to weigh the relative importance of the degrees of freedom or system states. Once it is determined which equations are to be eliminated, the remaining equations are modified to include an estimate of the effect of those that are removed. This process is discussed in most controls texts, and also in many texts on the finite-element method. Inman (1989) gives a very clear description and a brief example.

The differential equations governing a multi-degree-of-freedom mechanical system tend to have a predictable structure,

$$\mathbf{M}\ddot{\mathbf{x}}(t) + \mathbf{C}\dot{\mathbf{x}}(t) + \mathbf{K}\mathbf{x}(t) = \mathbf{f}(t), \tag{8.1}$$

where $\mathbf{M}$ and $\mathbf{K}$ are symmetric and positive definite and $\mathbf{C}$ is symmetric and positive semidefinite. There are certainly other forms of the equation of motion, such as those of gyroscopic and circulatory systems, but the structure of those forms is often similarly regular. It is natural to exploit any such features of the second-order governing equations in developing a first-order state-variable model. There has also been interest in developing control theory directly in terms of second-order plant equations, taking advantage of features such as symmetry of the coefficient matrices. See Skelton (1988) for a thorough treatment of this approach.

Particularly challenging from the standpoint of modeling are those mechanical systems with infinitely many degrees of freedom, that is, structures idealized as continua. These are governed by partial rather than ordinary differential equations, their stiffness, mass, and damping being represented by differential operators instead of matrices, and their response described by functions instead of vectors. (This excludes finite-element models, which are essentially discrete approximations to nominally continuous structures.) Part of the difficulty in dealing with continuous systems arises because the proper mathematical tools come from the field of functional analysis rather than from linear algebra, which is more familiar to most engineers. Although many of the concepts and methods from the linear algebra applicable in the analysis of discrete systems have analogs in functional analysis, it is often necessary when working with infinite sets of functions to explicitly establish mathematical structure and properties that can be taken for granted in the finite-dimensional case.

Even a structure modeled as a continuum must eventually be represented by a finite number of scalar differential equations in order for most types of structural dynamic and control calculations to be practicable. A common approach is to represent the response of a structure in an infinite series of the eigenfunctions of its mass and stiffness operators, that is, a series of the system's natural modes, truncating this series to a number of terms that yields a sufficiently accurate value of the response. This can lead directly to finite-dimensional matrix equations in which the generalized coordinates are modal quantities (i.e., natural frequencies, modal masses, and modal damping ratios). The coefficient matrices of these equations typically have properties like those mentioned above; in many cases the modal equations are decoupled, or coupled only through damping terms. However, an important difference between the matrix equation governing a finite-degree-of-freedom system and that obtained by truncating the modal series representation of a continuous system is that in the latter some degrees of freedom necessarily go unmodeled. If any of these degrees of freedom are important in the response of the structure or the performance of an active system of which it is a part, their omission may result in an inadequate model.

8.2.2 Control Strategies and Limitations

Assuming a model of reasonable size (dimension) and acceptable accuracy is available to represent a structure, we can turn to the calculation of a control law. In the simplest cases, it may be possible to proceed directly by using the tools described in Appendix B. More often, though, it will be expedient to consider the physics of the problem simultaneously with the mathematics. For example, negative feedback of position or rate is in many ways similar to adding stiffness or damping to a mechanical system. Part of the stiffness or damping in an active vibration absorber may be achieved through feedback, with the result that the absorber can be tuned by varying the feedback gains. Note, however, that if the sensed motion of one mass is used in computing the control force to be applied to another mass to which the first is not directly (mechanically) connected, the coupling in the closed-loop system will be qualitatively different from that in the passive, open-loop structure. This is not necessarily undesirable, but it serves to illustrate that the dynamics of the active system can quickly become much more complex than those of the structure alone. This complexity extends to such fundamental matters as stability; further, it can introduce a very large number of parameters with less obvious physical meanings than "stiffness" and "damping," for example, if every mass in a MDOF system is subject to control forces depending on the motion of every other mass.

When a structure consists of lumped masses and springs, it is possible at least in principle to dedicate a sensor and an actuator to each degree of freedom. Possible controller structures then range from the strictly local, where the signal driving each actuator is derived from the output of only the corresponding sensor, to the configuration in which each actuator signal is computed using information from every sensor. The latter introduces new coupling into the equations of motion, and may affect the symmetry of the coefficient matrices as well. These effects have been investigated extensively, and some (very stringent) conditions under which they are minimized are known; see Inman (1989) and Meirovitch (1990) for summaries of these results.

Many of the same phenomena are observed when spatially discrete ("point") sensors and actuators are used to control a continuous structure. In addition, it is more likely that the sensors and actuators will be at different locations on the structure (although the configuration in which they are collocated is an important special case and permits some simplifications). Because the plant has infinitely many degrees of freedom and thus infinitely many states, the effects of unmodeled dynamics must be addressed, for example, by ensuring that all significantly contributing modes are included in any finite-dimensional model. It is also often necessary to represent the control forces (actuator outputs) in the modal or state coordinates, and then to construct physical control forces or actuator commands from the results of modal or state-space control law computations. These problems have received a great deal of attention in the literature; the reader is referred to the texts and survey articles cited at the end of this chapter for details.

Materials like the piezoelectric crystals and shape memory alloys discussed in Chapters 2 and 3 of this book offer means of constructing sensors and actuators that work over a finite area of a distributed structure rather than at a discrete point. Because these devices integrate (in the sense of calculus) over that region of the structure to which they are attached or in which they are embedded, they do not measure or respond to spatial information in the way that is often desired for use in spatially continuous structural control. For example, this unavoidable spatial integration can render sensors ineffective in detecting some structural modes, depending on the shape of the mode and the placement of the sensor. Measuring response as a function of space as well as time is more challenging, although it is possible through techniques such as holography. The fiber-optic sensors reviewed in Chapter 7 are among the most promising candidates for compact, lightweight, inexpensive distributed response measurement.

8.3 Active Structures in Practice

In this section, we shall consider some examples of smart structures that illustrate the potential of integrating sensing, actuation and control into a structural system. Note that even in such proof-of-concept work a multidisciplinary approach is generally necessary, and how the initial problem statement has ramifications throughout the design, analysis, and operation of even a simple smart structure. Additional representative applications may be found in the references given at the end of this chapter, and in Chopra (1996). The reader interested in aerospace technology is encouraged to refer to Wie (1988) and Frank et al. (1994), who present broad, accessible treatments of the development of active smart structures for space flight.

8.3.1 Systems Using SMA Actuators

Summarized here are examples of smart structures control as approached from the point of view of applied mechanics or that of automatic control. Both cited papers give clear descriptions of the analytical and experimental procedures that were followed, but the first is concerned primarily with mechanics while the second addresses in detail the control of a simple laboratory structure.

The use of a simple feedback control scheme together with embedded shape memory alloy actuators similar to that described in Section 3.6 is discussed in Choi and Lee (1998). In the experiments reported there, SMA wire actuators were embedded directly in composite beam specimens, which were then arranged with fixed boundary conditions at each end and subjected to a controllable axial load P. The plane of the SMA wires was offset from the neutral axis of the beam, with the result that when the wires were energized by electrical resistance heating they had the effect of applying a distributed moment, M_{SMA}, to the beam. In each test, the beam was given a prescribed initial deflection at its midpoint, so that the axial compression produced a moment M_P. The lateral deflection δ of the beam at its midpoint was sensed during loading and used to control the switching of electrical current which heated the SMA wires.

The buckling response of this column was analyzed by using both the conventional Euler method and a "cut and paste" method in which the stress and displacement fields in the SMA wires and the surrounding composite beam were considered separately, with similar results. It was predicted, and verified experimentally, that the control moment M_{SMA} could be used to increase the critical load of the column. In some cases, the moment produced by the SMA wires was so great that the initial displacement could be entirely recovered, followed by snap-through of the beam. The results of varying the initial displacement and the displacement threshold at which the control current was switched on are discussed in detail by Choi and Lee (1998).

By comparison, Rhee and Koval (1993) treated simpler structures (a cantilever beam and a discrete spring-mass system) but devoted a good deal of effort to modeling them for control purposes. The SMA wires used as actuators in both structures were considered alone, and representations in the form of transfer functions and state-variable models were obtained for use respectively in the design of classical and robust control laws for the closed-loop systems of which they were part. This modeling is typical of what must be done in applying control techniques to smart structures, and is sketched here as an example of the procedures commonly used.

The electrically heated SMA wire actuators were arranged to impart force or moment by their constrained contraction. These wires were not embedded, but stood off from the structures to which they were attached. Three lengths of wire were used – 24 in., 42-in., and 54-in.; all were characterized in the same way for later use in control design. The wires were given a 3% prestrain, constrained at their ends, and put through a heating-cooling cycle, during which the force they developed was recorded. A first-order model of the form

$$\tau \dot{F} + F = a_b P \tag{8.2}$$

was fit to the measured force response, where

$$F = \text{force,} \tag{8.3}$$

$$\tau = \text{time constant,} \tag{8.4}$$

$$a_b = \text{slope of force vs. power curve,} \tag{8.5}$$

$$P = \frac{V^2}{R} = \text{power supplied to wire,} \tag{8.6}$$

$$V = \text{voltage,} \tag{8.7}$$

$$R = \text{resistance of wire.} \tag{8.8}$$

The time constant $\tau = 5$ s of the exponential force response was determined experimentally for a typical driving power P in the range of 2 to 4 W. In the steady state, the force versus power curve was found to have a slope of approximately 1.25 lb/W. The voltage applied to the wire was linearly related to the voltage supplied by the PC-based controller, $V = 2.2V_{\mathrm{PC}}$, and the 42-in. wire was found to have a resistance of 18 Ω. When this information was used to fix the constants in the model equation, it became

$$5\dot{F} + F = 0.336V_{\mathrm{PC}}^2, \tag{8.9}$$

and linearizing this about a nominal V_{PC} of 2.5 V yielded

$$5\dot{f} + f = 1.68V_{\mathrm{PC}}. \tag{8.10}$$

The corresponding transfer function of this actuator is

$$G_a(s) = \frac{0.336}{s + 0.2}. \tag{8.11}$$

After testing with 5 V applied to the wire, this was revised to

$$G_a(s) = \frac{0.15}{s + 0.2}, \tag{8.12}$$

which was used in the subsequent design work. Similarly, the transfer function of the 54-in. wire was found to be

$$G_a(s) = \frac{0.27}{s + 0.2}, \tag{8.13}$$

and that of the 24-in. wire,

$$G_a(s) = \frac{0.302}{s + 0.2}. \tag{8.14}$$

No forced convection or other active cooling was used. Not unexpectedly, the bandwidth of the SMA actuators was limited by heat transfer. The 54-in. wire, for example, was found to be limited to use at frequencies below 2 Hz.

The first structure considered was an aluminum cantilevered beam, of dimensions 47 in. × 6 in. × 0.125 in. Two 42-in. wires were attached to each side of the beam as actuators, as shown in Fig. 8.1, and a tip mass of up to 6.44 lb was added to alter the fundamental natural frequency of vibration of the beam. The motion of the beam was sensed by a strain gage mounted near its root.

The beam was analyzed by using the finite-element method, and both three- and ten-element discretizations were found to give similar results for the lower frequencies. In

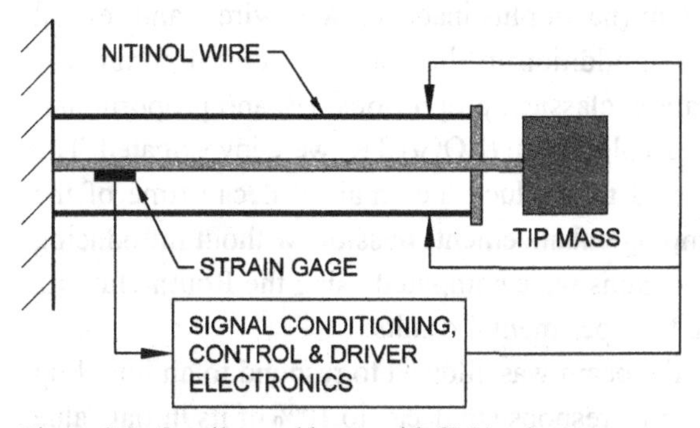

Figure 8.1. Cantilevered beam with SMA wire actuators studied by Rhee and Koval (1993).

order to use classical control design techniques, a single-input, single-output (SISO) model was required, limiting the analysis to a single beam mode. The transfer function of the beam was derived starting with the differential equation of its fundamental mode, obtained from the finite-element analysis as

$$\ddot{q}_1 + 0.0235\dot{q}_1 + 15.31q_1 = 1.684m_a(t). \tag{8.15}$$

Here, q_1 is a generalized coordinate and $m_a(t)$ is the moment applied to the beam by the SMA actuators; this equation also reflects the experimentally measured beam damping ratio of 0.30%. The axial compression of the beam caused by the actuator force was neglected, and the control moment was given in terms of the actuator force $f(t)$ and the distance of the actuators from the beam's neutral axis, 1.5 in., as simply

$$m_a(t) = 1.5f(t). \tag{8.16}$$

To obtain the beam equation of motion in terms of a physical, rather than a generalized, coordinate, it was rewritten in terms of the tip displacement y_T. In this case, $y_T = q_1$, and therefore

$$\ddot{y}_T + 0.0235\dot{y}_T + 15.31y_T = 2.526f(t). \tag{8.17}$$

The strain gage output voltage v was proportional to the beam tip displacement, and using the experimentally determined sensitivity of 0.0624 V/lb the equation of motion can be written in terms of this output quantity as

$$\ddot{v} + 0.0235\dot{v} + 15.31v = 0.0624f(t). \tag{8.18}$$

This differential equation represents a relationship between the beam's input, the force $f(t)$, and its output, the strain gage voltage v. The desired beam transfer function can therefore be obtained by Laplace transformation, which yields

$$G_p(s) = \frac{0.0624}{s^2 + 0.0235s + 15.31}. \tag{8.19}$$

The corresponding frequency response was compared with that obtained experimentally and found to agree acceptably.

Simulations of the open-loop system (beam plus inactive SMA wires) and several control algorithms were carried out. In addition to the "on-off" control commonly reported in the smart structures literature, classical proportional (P) and proportional-plus-integral (PI) designs and one robust algorithm (LQG/LTR) were investigated. The P and PI controller gains were selected to produce the smallest decay time of the free-vibration response (greatest damping enhancement) possible without introducing instability. The stable region(s) for the gains were computed using the Routh-Hurwitz criterion and found to be comparable to experimental results.

In experiments on these systems, the beam was allowed to respond to an initial tip displacement, and the time required for its response to decay to 10% of its initial value was used as a measure of damping performance. The addition of a tip mass of 6.44 lb

Table 8.1. Comparison of Control Algorithms for Cantilevered Beam (10% Settling Times, Seconds) (Rhee and Koval, 1993)

Controller	Simulation	Experiment
Open-loop	—	220
On-off	37	38
P	22	24
PI	36	34
LQG/LTR	19	17

produced a fundamental frequency of 0.623 Hz. The results of these experiments are summarized in Table 8.1, where it may be seen that agreement between simulations and experiments is generally quite good. All of the closed-loop algorithms are successful in that they increased the damping of the beam's fundamental mode of vibration significantly, but the robust controller performs somewhat better than the best classical controller (P).

A similar process of modeling and controller design was carried out for the second structure studied, a three-degree-of-freedom spring-mass system shown schematically in Fig. 8.2. The most important difference from the case of the cantilevered beam is that this system cannot be simplified to a SISO model, but is treated as a multi-input, multi-output (MIMO) plant. Consequently, of the controller types previously considered, only the robust (LQG/LTR) algorithm is applicable here.

Two sets of actuator wires were used, 54-in. wires running top-to-bottom and 24 in. wires crossed within one story of the structure. Strain gages bonded to the top and bottom springs were used to sense the motion of the masses.

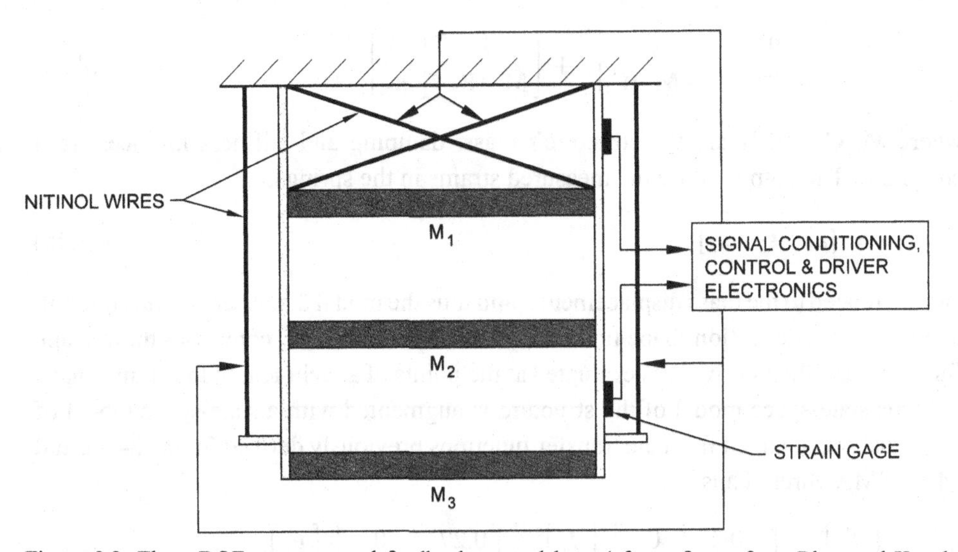

Figure 8.2. Three-DOF structure and feedback control loop (after a figure from Rhee and Koval, 1993).

Finite-element analysis of this structure revealed three modes within the bandwidth of the actuators (below 3 Hz). The fourth and higher natural frequencies were all greater than 35 Hz, and corresponded to spring resonances and other motions in which the masses participated only slightly. Therefore, three modes were used in formulating a state-variable model of the structure, based on the FEA results plus a 0.3% damping ratio. This led to an equation of the form (8.1),

$$\begin{bmatrix} 0.087 & 0 & 0 \\ 0 & 0.108 & 0 \\ 0 & 0 & 0.091 \end{bmatrix} \begin{bmatrix} \ddot{q}_1 \\ \ddot{q}_2 \\ \ddot{q}_3 \end{bmatrix} + \begin{bmatrix} 0.002 & 0 & 0 \\ 0 & 0.0051 & 0 \\ 0 & 0 & 0.0072 \end{bmatrix} \begin{bmatrix} \dot{q}_1 \\ \dot{q}_2 \\ \dot{q}_3 \end{bmatrix}$$

$$+ \begin{bmatrix} 1.9 & 0 & 0 \\ 0 & 15.8 & 0 \\ 0 & 0 & 28 \end{bmatrix} \begin{bmatrix} q_1 \\ q_2 \\ q_3 \end{bmatrix} = \begin{bmatrix} -0.45 & -0.017 \\ -1.0 & 0.108 \\ 0.654 & 0.129 \end{bmatrix} \begin{bmatrix} f_2 \\ m_5 \end{bmatrix}, \tag{8.20}$$

where the q_i are again generalized coordinates, f_2 is the effective transverse force from the crossed SMA wires, and m_5 is the moment applied to the structure by the wires running parallel to the frame.

Corresponding state-space equations

$$\dot{\mathbf{x}} = \mathbf{A}_s \mathbf{x} + \mathbf{B}_s \mathbf{u} \tag{8.21}$$

$$\mathbf{y} = \mathbf{C}_s \mathbf{x} \tag{8.22}$$

are obtained by defining the state vector

$$\mathbf{x} = [q_1 \quad q_2 \quad q_3 \quad \dot{q}_1 \quad \dot{q}_2 \quad \dot{q}_3]^T. \tag{8.23}$$

Then

$$\dot{\mathbf{x}} = \begin{bmatrix} \mathbf{0} & \mathbf{I} \\ -\mathbf{M}^{-1}\mathbf{K} & -\mathbf{M}^{-1}\mathbf{C} \end{bmatrix} \mathbf{x} + \begin{bmatrix} \mathbf{0} \\ \mathbf{M}^{-1}\mathbf{B}_s \end{bmatrix} \begin{bmatrix} f_2 \\ m_5 \end{bmatrix}, \tag{8.24}$$

where **M**, **C**, and **K** are the structure's mass, damping and stiffness matrices from eq. (8.20). The outputs **y** are the measured strains in the springs,

$$\mathbf{y} = [\varepsilon_1 \quad \varepsilon_2 \quad \varepsilon_3]^T, \tag{8.25}$$

and are related to the mass displacements (and thus the modal displacements q_i) through assumed beam deflection shape functions. The output matrix $\mathbf{C}_s$ comprises these shape functions and their derivatives evaluated at the points of attachment of the strain gages.

This state-space model of the structure is augmented with a state-space model of the actuators, obtained from the transfer functions previously derived for the 24-in. and 54-in. SMA wires. Thus,

$$\begin{bmatrix} \dot{f}_s \\ \dot{f}_1 \end{bmatrix} = \begin{bmatrix} -0.2 & 0 \\ 0 & -0.2 \end{bmatrix} \begin{bmatrix} f_s \\ f_1 \end{bmatrix} + \begin{bmatrix} 0.27 & 0 \\ 0 & 0.302 \end{bmatrix} \begin{bmatrix} v_1 \\ v_2 \end{bmatrix} \tag{8.26}$$

or

$$\dot{\mathbf{f}}_a = \mathbf{A}_a\mathbf{f}_a + \mathbf{B}_a\mathbf{v}. \tag{8.27}$$

The force and moment applied to the structure are related to the actuator forces by

$$\begin{bmatrix} f_2 \\ m_5 \end{bmatrix} = \begin{bmatrix} \sin\theta & 0 \\ 0 & a \end{bmatrix} \begin{bmatrix} f_s \\ f_1 \end{bmatrix}, \tag{8.28}$$

$$= \mathbf{C}_a\mathbf{f}_a \tag{8.29}$$

where θ is the inclination angle of the crossed wires and a is the stand-off distance of the axial wires from the structure. When these actuator dynamics are combined with the structural model above, the augmented state-space model is

$$\begin{bmatrix} \dot{\mathbf{x}} \\ \dot{\mathbf{f}}_a \end{bmatrix} = \begin{bmatrix} \mathbf{A}_s & \mathbf{B}_s\mathbf{C}_s \\ \mathbf{0} & \mathbf{A}_a \end{bmatrix} \begin{bmatrix} x \\ f_a \end{bmatrix} + \begin{bmatrix} \mathbf{0} \\ \mathbf{B}_a \end{bmatrix} \mathbf{v}, \tag{8.30}$$

$$\mathbf{y} = [\mathbf{C}_s \quad \mathbf{0}] \begin{bmatrix} \mathbf{x} \\ \mathbf{f}_a \end{bmatrix}. \tag{8.31}$$

A robust controller was designed to control the first two modes, again with the goal of enhancing the damping of the structure. The time for the free vibration response to decay to 10% of its initial value was compared experimentally for the open-loop structure and the closed-loop system, and a significant improvement in damping was observed. Simulation results also agreed well with experiment. The robustness of the controller against modeling error was tested by adding a 4.4 lb weight to the third mass, under which condition the performance of the controller was somewhat degraded but the system remained stable. Details may be found in Rhee and Koval (1993).

8.3.2 Systems Using PZT Sensors and Actuators

The first paper reviewed here (Finefield et al., 1992) illustrates the opportunities offered by smart structures for implementing control schemes that would be prohibitively cumbersome or complex if carried out by using other technologies. Two control loops, an analog feedback effective in countering broadband excitations and an adaptive least mean square (LMS) feedforward tuned for a harmonic component of the disturbance, were used together and found to perform better than either control alone in suppressing structural vibration and acoustic radiation.

The structure examined was a thick-walled aluminum cylinder, to which were bonded piezoelectric film (PVDF) strain and strain-rate sensors and piezoelectric (PZT) induced-strain actuators. A conventional strain gage and an accelerometer were used to take additional data on the vibration of the cylinder. The cylinder was driven at a single point by an electromechanical shaker. The amount and type of response information available from this array of sensors, and the ease of applying control forces via the PZT actuators, far exceeds that typically taken with discrete, nonintegrated devices. This paper focused on the establishment of the proper functioning of this instrumentation and on the simultaneous application of the analog and adaptive digital controls described, carried out in a sequence of six experiments:

1. Verification of the relationship between sensor strain and the strain rate signal
2. Using strain rate feedback to add damping to the structure
3. Suppressing vibration with analog feedback control
4. Suppressing vibration with digital adaptive feedforward control
5. Suppressing vibration with a combination of these two controls
6. Reducing acoustic emission from the cylinder

The growth in complexity in the experimental setups and data processing is documented and discussed, followed by a presentation of experimental findings. The control algorithms are shown to work individually and in combination. In the latter case, the broadband vibration suppression provided by the analog control loop was found to aid the operation of the narrowband adaptive LMS controller. This study demonstrates that as basic smart-structures technologies, such as PZT and PVDF sensors and actuators, mature, the complexity and performance of feasible systems quickly grows.

The details of building general-purpose structural control devices using PZT sensors and actuators were taken up in Bronowicki et al. (1994), motivated by the observation that surface-mounted PZT sensors and actuators are applicable in many applications where the through-thickness response of the structure is not a concern, while the associated electronics are rather generic. Advantages of ease in retrofitting such systems, and of simplicity and economy of separating the design, construction, and testing of the control system from that of the base structure, are also cited. (It will be recognized that a greater degree of integration of the components of a smart structure is possible when all its parts are designed and fabricated concurrently, but the costs of realizing this potential can be prohibitive.) Bronowicki et al. (1994) present the details of the physical configurations of PZT patches on a model structure, then describe the construction and performance of two controllers that might be used with this type of installation. We sketch some of their findings here, concentrating on the structural aspects of these systems; much more detail on the controller electronics and programming may be found in the cited paper.

Sensor and actuator elements consisting of collocated PZT patches were bonded to a "conical stub," a graphite composite shell 56 in. in length and tapering from a diameter of 3.5 in. at its free end to 1.8 in. at the end where it was mounted as a cantilever. Two sensor-actuator patches were bonded to opposite sides near the small end. Each of these comprised an 8 in. × 1.25 in. actuator patch and a 1.5 in. × 1 in. sensor collocated in the center of the actuator region. Two "nearly collocated" sensors were placed 1 in. beyond the ends of the actuator region. Altogether, these patches added approximately 5% to the weight of the structure.

An external SISO analog controller consisting of a charge amplifier, a compensator filter with jumper-selectable lowpass and bandpass characteristics, and a power amplifier was connected to the sensor and actuator patches by standard cabling. The bandwidth over which the structure's damping was augmented could be varied by tuning the filters in the feedback loop. The controller could be configured to implement integral feedback control if the highpass filter cutoff frequency was set below the natural frequencies of structural vibration, while a narrower bandwidth was used with rate-feedback control. Following quantification of the generally negligible effects of electrical noise from the power supplies, this analog controller was demonstrated to yield modal damping levels of up to 100% in simple, cantilever structures, and up to 20% when used with similar PZT sensor-actuator patches applied to larger, more complex structures. Control (reduction in resonant response levels) was achieved over a bandwidth of 10 to 200 Hz.

A digital controller with two inputs and two outputs, based on the Texas Instruments TMS320C30 digital signal processor and including on the same board the requisite analog sensor and actuator electronics, was designed and fabricated for use with the same type of PZT patches. The circuitry included analog lowpass anti-aliasing filters, off-chip ROM for boot code, and off-chip RAM for storage of code implementing large control algorithms. The frequency response of this controller was found to be flat out to 10 kHz, where the effect of the (adjustable) anti-aliasing filters began to be seen. This configuration of the hardware was found to be capable of controlling structural modes up to 2 kHz. In the example application of the digital controller to a cantilevered active structural member with primary modes at frequencies of 7, 70, and 200 Hz, the authors report that "closing the loop provided significant damping in all three modes of vibration, with 100% damping obtained in the first mode." Using a more sophisticated adaptive LMS algorithm, the same structure was controlled with a noncollocated accelerometer and PZT actuator. The response of the tip of the cantilever was reduced by a factor of 30.

The development of generic sensor-actuator patches and controller modules can obviously simplify the construction of active smart structures, and has the practical benefit of reducing design effort, fabrication costs, and testing requirements. The trade-off, of course, is between these savings and the possibly superior performance of a custom control system tailored for a single application. But just as standardization of raw materials such as fasteners and composite prepreg strips has allowed the realization of many novel passive structures at acceptable costs, the availability of modular control components can be expected to stimulate the use of active control in applications where it would otherwise be impractical.

8.4 Summary

Throughout this chapter we have been concerned with topics that conventionally belong to the subject of control theory, and with some of the features and complications that distinguish the field of structural control. It will be apparent that some of the devices discussed in this book fit into this framework better than do others – for example, a small, shaped PZT patch driven by a controlled voltage may be well approximated as a linear actuator driving a large continuous structure at a single point, but a long, embedded SMA wire responding to electrical resistance heating cannot be so simply idealized. ERF and MRF fluids and some uses of SMAs can radically alter a structure's properties in response to a command signal without directly exerting a control force. This type of semi-active control can be quite effective and efficient. Applications are often described in the literature pertaining to a particular technology, e.g., SMA, as well as in that devoted to structural control.

Because many (if not most) smart structures depend on open- or closed-loop active control, the designer or analyst must be familiar with a variety of topics in the fields of structural dynamics and automatic control. These include not only linear systems theory and classical and modern control system design, but also the complexities of structural models. Although it is often tempting to view the problem from the standpoint of structural dynamics or that of control theory, the best results will be obtained when both these aspects are considered simultaneously.

BIBLIOGRAPHY

Abdel-Rohman, M., and Horst H. E. Leipholz. 1984. Optimal feedback control of elastic, distributed parameter structures. *Computers & Structures* 19(5/6):801–805.

Balas, M. J. 1978. Active control of flexible systems. *Journal of Optimization Theory and Applications* 25(3):415–436.

Banks, H. T. ed. 1992. *Control and Estimation in Distributed Parameter Systems*. Philadelphia: Society for Industrial and Applied Mathematics.

Banks, H. T., R. C. Smith, and Y. Wang. 1994. Vibration suppression with approximate finite dimensional compensators for distributed systems: Computational methods and experimental results. In *Proceedings of the Second International Conference on Intelligent Materials*, 140–154, Colonial Williamsburg, Virginia.

Bronowicki, A., J. Innis, S. Casteel, G. Dvorsky, O. Alvarez, and E. Rohleen. 1994. Active vibration suppression using modular elements. In *Proceedings of the SPIE Conference on Smart Structures and Intelligent Systems* 2190:717–728, Orlando, Florida: SPIE.

Choi, S., and J. Ju Lee. 1998. The shape control of a composite beam with embedded shape memory alloy wire actuators. *Smart Materials and Structures* 7:759–770.

Chopra, I. 1996. Review of current status of smart structures and integrated systems. In *Proceedings of the SPIE conference on smart structures and integrated systems* 2717: 20–62, San Diego: SPIE.

Finefield, J. K., H. Sumali, and H. H. Cudney. 1992. Combined feedback and adaptive digital control of cylinder vibrations using smart materials. In *Proceedings of the 31st Conference on Decision and Control* TM3–16:20:1809–1814, Tucson, Arizona: IEEE.

Frank, G. J., S. E. Olson, and M. L. Drake. 1994. Jitter suppression experiment active damping system design considerations. In *Proceedings of the SPIE Conference on Smart Structures and Intelligent Systems* 2190:802–812, Orlando, Florida: SPIE.

Hale, A. L., and G. A. Rahn. 1984. Robust control of self-adjoint distributed-parameter systems. *AIAA Journal of Guidance* 7(3):265–273.

Inman, D. J. 1989. *Vibration: With Control, Measurement and Stability*. Englewood Cliffs, NJ: Prentice Hall.

Junkins, J. L., and Y. Kim. 1993. *Introduction to Dynamics and Control of Flexible Structures*. Washington, DC: American Institute of Aeronautics and Astronautics.

Kamada, T., T. Fujita, T. Hatayama, T. Arikabe, N. Murai, S. Aizawa, and K. Tohyama. 1997. Active vibrations control of frame structures with smart structures using piezoelectric actuators (vibration control by control of bending moments of columns). *Smart Materials and Structures* 6:448–456.

Khorrami, F., N. Das, and S. Nourbakhsh. 1997. Microstrip antennas for wireless communication in smart structures and active damping. In *Proceedings of the SPIE Conference on Smart Electronics and MEMS* 3046:84–93, San Diego: SPIE.

Kolmogorov, A. N., and S. V. Fomin. 1975. *Introductory Real Analysis*. New York: Dover Publications, Inc.

Lam, K. Y., and T. Y. Ng. 1999. Active control of composite plates with integrated piezoelectric sensors and actuators under various dynamic loading conditions. *Smart Materials and Structures* 8:223–237.

Leng, J., and A. Asundi. 1999. Active vibration control system of smart structures based on fos and er actuator. *Smart Materials and Structures* 8:252–256.

Leo, D., and D. Inman. 1994. Convex controller design for vibration suppression of a flexible antenna. In *Proceedings of the Second International Conference on Intelligent Materials*, 816–827, Colonial Williamsburg, VA: Technomic Publishing Co., Inc.

Main, J. A., G. Nelson, and J. Martin. 1998. Electron gun control of smart materials. In *Proceedings of the SPIE Conference on Smart Structures and Integrated Systems* 3329:688–693, San Diego: SPIE.

McClamroch, N. H., D. S. Ortiz, H. P. Gavin, and R. D. Hanson. 1994. Electrorheological dampers and semi-active structural control. In *Proceedings of the 33rd Conference on Decision and Control* WA–4 10:20:97–102, Lake Buena Vista, FL: IEEE.

Meirovitch, L., H. Baruh, and H. Öz. 1993. A comparison of control techniques for large flexible systems. *AIAA Journal of Guidance* 6(4):302–310.

Meirovitch, L. 1990. *Dynamics and Control of Structures*. New York: John Wiley & Sons. 1990.

Miller, R. K., S. F. Masri, T. J. Dehghanyar, and T. K. Caughey. 1988. Active vibration control of large civil structures. *ASCE Journal of Engineering Mechanics* 14(9):1542–1570.

Park, Y. K., and S.-B. Choi. 1997. Dynamic modeling and shape control of a flexible plate containing electrorheological fluids. In *Proceedings of the SPIE Conference on Smart Structures and Integrated Systems* 3041:528–535, San Diego: SPIE.

Pota, H. R., T. E. Alberts, and I. R. Petersen. 1993. H^{∞} control of flexible slewing link with active damping. In *Proceedings of the SPIE Conference on Smart Structures and Intelligent Systems* 1917:59–70, Alburquerque, NM: SPIE.

Reinhorn, A. M., and G. D. Manolis. 1989. Recent advances in structural control. *Shock and Vibration Digest* 21(1):3–8.

Rhee, S. W., and L. R. Koval. 1993. Comparison of classical with robust control for sma smart structures. *Smart Materials and Structures* 2:162–171.

Skelton, R. E. 1988. *Dynamic systems control: Linear systems analysis and synthesis*. New York: John Wiley & Sons. 1988.

Soong, T. T. 1990. *Active Structural Control: Theory and Practice*. Essex, England: Longman Scientific and Technical.

Spencer, B. F., Jr. and M. K. Sain. 1997. Controlling buildings: A new frontier in feedback. *IEEE Control Systems* 17(6):19–35.

Tsai, M. S., and K. W. Wang. 1996. Control of a ring structure with multiple active-passive hybrid piezoelectrical networks. *Smart Materials and Structures* 5:695–703.

Wie, B. 1988. Active vibration control synthesis for the control of flexible structures mast flight system. *AIAA Journal of Guidance, Control and Dynamics* 11(3):271–277.

Young, A. J., and C. H. Hansen. 1994. Control of flexural vibration in a beam using a piezoceramic actuator and an angle stiffener. *Journal of Intelligent Material Systems and Structures* 5:536–549.

PROBLEMS

1. A single-degree-of-freedom mechanical oscillator consists of a mass $m = 1$ kg, a spring with stiffness $k = 100$ N/m, and a dashpot with coefficient selected so that the damping ratio $\zeta = 2\%$. Acting on the mass are an external load $f(t)$ and a control force $u(t)$, both taken to be positive in the direction of positive displacement $x(t)$.

(a) Write the differential equation of motion of this system in the time domain.

(b) Transform the equation obtained in part (a) to the complex frequency domain. Include any terms due to potentially nonzero initial conditions $x(0) = x_0$, $\dot{x}(0) = \dot{x}_0$.

2. For the system of Problem 1, let the control law be given by negative feedback of the velocity and displacement of the mass,

$$u(t) = -g_1\dot{x}(t) - g_2 x(t). \tag{8.32}$$

(a) Draw a block diagram of the closed-loop system in the frequency domain.

(b) Calculate the closed-loop transfer function and the characteristic equation. Show the system is stable when $g_1 = g_2 = 0$.

(c) Find the gains g_1 and g_2 so that the active system has a natural frequency of 11 rad/s and a damping ratio of 5%.

(d) With the gains at the values calculated in part (c), use the final value theorem to find the response of the system to the step load $f(t) = 2H(0)$ as $t \to \infty$. ($H(t)$ is the Heaviside step function, 0 for $t < 0$ and 1 otherwise.)

(e) Determine the range of the feedback gains within which the system's step response is stable and oscillatory.

3. Consider the system of Problem 1 with an unspecified control $u(t)$.

(a) Write the state equation of this system in terms of the variables $x_1 = x$, $x_2 = \dot{x}$.

(b) Specialize this result to the control law given in Problem 2.

(c) Show that the characteristic equation obtained from the matrix $\mathbf{A}$ has the same roots as that found in Problem 2(b).

4. Simulate the response of the system of Problem 3(b) to the step load of Problem 2(d). Assume zero initial conditions ($x_0 = \dot{x}_0 = 0$).

(a) Plot the evolution of the state variables for the duration of the transient response.

(b) Compute and plot the energy dissipated in the dashpot as a function of time.

5. Consider the system of Problem 3(b) and let the applied load be a unit impulse occurring at time $t = 0$, that is, $f(t) = \delta(0)$. Assume zero initial conditions.

(a) Formulate the linear quadratic regulator problem for this system (let $t_f = \infty$).

(b) Simulate numerically the system response, making note of any further assumptions or approximations needed to facilitate the computation. Demonstrate the effect of the choice of weighting matrices on the optimal response and the value of the cost function.

6. Investigate (on physical grounds) the notions of controllability and observability in a 2-DOF mechanical system under the restriction that one sensor and one actuator are attached to either mass, in any combination.

7. Consider a simply supported beam of length l to which is bonded a PZT patch of the type described in Chapter 2. The patch is of length a along the beam axis and is the full width of the beam. It is attached to the top surface so as to act as a sensor responding to the curvature of the beam.

(a) Assume the natural frequencies and modes of this structure are the same as those of the beam alone. If the patch is located at the middle of the span, $x = l/2$, to which mode(s) will it be least sensitive?

(b) What do the results of part (a) suggest about the effectiveness of an embedded SMA actuator running the full length of the beam in controlling motion in the second vibration mode? Does this apply if the beam boundary conditions are unsymmetric, for example, if the beam is cantilevered?

8. An active vibration absorber of the type shown in Fig. 5.16 has a mass 5% of that of the primary structure to which it is attached, that is, $m_a = 0.05m$.

(a) For fixed values of k_a and c_a, determine the allowable ranges of the displacement and velocity feedback gains g_d, g_v, where

$$f_a = -g_v(\dot{x}_a - \dot{x}) - g_d(x_a - x). \tag{8.33}$$

That is, find the values of these gains that result in a stable, oscillatory response of the absorber mass.

(b) Calculate those values of the gains that result in the most effective suppression of harmonic motion of the primary mass when it is driven by a load of frequency ω.

(c) Investigate the effect of the feedback gains on the response of both masses to a step load applied to the primary mass. Compare the results of analytical methods from classical control theory with direct numerical simulations.

9

Biomimetics

9.1 Introduction

> The X–29 and HI-MAT use on-board computers to provide greater flexibility. Designers hope that someday these techniques will give man-made flying machines the agility of a dragonfly, which can hover and change direction almost instantly in its search for food. (Footnote on a display at the National Air & Space Museum, Washington, D.C., August, 1990)

In this concluding chapter, we introduce the science and technology of certain structural systems in nature. Natural structures are smart structures par excellence inasmuch as they are the result of a continuous process of optimization over millennia. The field of biomimetics is at the intersection of biology and engineering and is dedicated to understanding strategies with which nature builds its structures and determining the extent to which some design principles may be incorporated in the design of man-made structures.

As emphasized in all the earlier chapters, future technological advances are dependent almost entirely on inventing new materials with engineered properties that permit design and development of smart structures. The prospect of mimicking biological synthesis in producing man-made materials has generally appeared to be an impossible task. In recent years, however, revolutionary advances in our ability to probe the fabric of materials down to their atomic structures and in the processing control of advanced materials' microstructure have fueled a renewed interest in imitating natural processes, initially in the laboratory and ultimately at the industrial scale. The engineering and scientific communities have a fundamental role to play in this new and promising endeavor.

The basic characteristics of natural structures, such as efficiency, precision, self-repair, and durability, continue to fascinate designers of engineering structures. The aerial acrobatics of birds, the ease with which the remarkably nimble dragonfly moves and maneuvers through the air, the precision with which bats sense, pinpoint, and capture their prey, and the meticulous manner in which strong and tough shells of mollusks are assembled from relatively brittle materials are but a few examples of the marvel of biological engineering and manufacturing. These examples, among others, can serve as excellent models for exploiting steady and unsteady flows, controlling motion, and developing superior material/structure concepts appropriate to the increasingly demanding performance requirements of flight vehicles of the next century. An overview of the subject along with a more complete list of references is found in articles prepared by the senior author (Srinivasan et al., 1991, 1996).

9.2 Characteristics of Natural Structures

There is no pretense that we can even begin to cover all the fascinating characteristics of naturally evolved materials and structures. Our purpose is only to hint at the ingenuity of nature by examining a few selected features of certain natural systems.

1. *Multifunctionality*

   This characteristic appears to be a common feature of most, if not all, biological systems and represents an ingenious "design" of the structure in which the individual components participate in more than one function. Nature accomplishes this through integration of and balance among material capabilities and structural and functional requirements.

   Gordon (1988) cites the example of sharp spines on the backs of hedgehogs. These spines are obvious protection for the animal from its predators. In addition, when the hedgehog descends from heights by rolling or falling, the spines cushion the impact by undergoing Eulerian buckling deformation (Vincent and Owers, 1986). Hadley (1986) in his discussion of arthropod cuticle identifies the principal function of the material to be that of limiting the loss of water from the animal. In addition, the cuticle provides (a) a support structure, (b) protection against the environment, (c) attachment locations for muscles, and (d) optical reflectance for camouflage and behavioral signalling. It is well known that roots of plants and trees not only anchor the structure to the ground but also serve as conduits for intake of water and transfer of needed nutrients.

   Lest we conclude that this feature is unique only to natural systems, let us not forget that, to the extent possible, engineering structures are also designed to be multifunctional. For example, airplane wings not only provide lift, which is their primary function, but also store fuel.

2. *Hierarchical Organization as a Basis for Structural Integrity*

   A study of natural materials/structures such as tendon, bone, wood, and shells reveals that a living organism builds the material and develops its architecture starting at levels well below that of the living cells. Tendon, for example, which serves as a link between muscle and bone, has an extremely intricate and

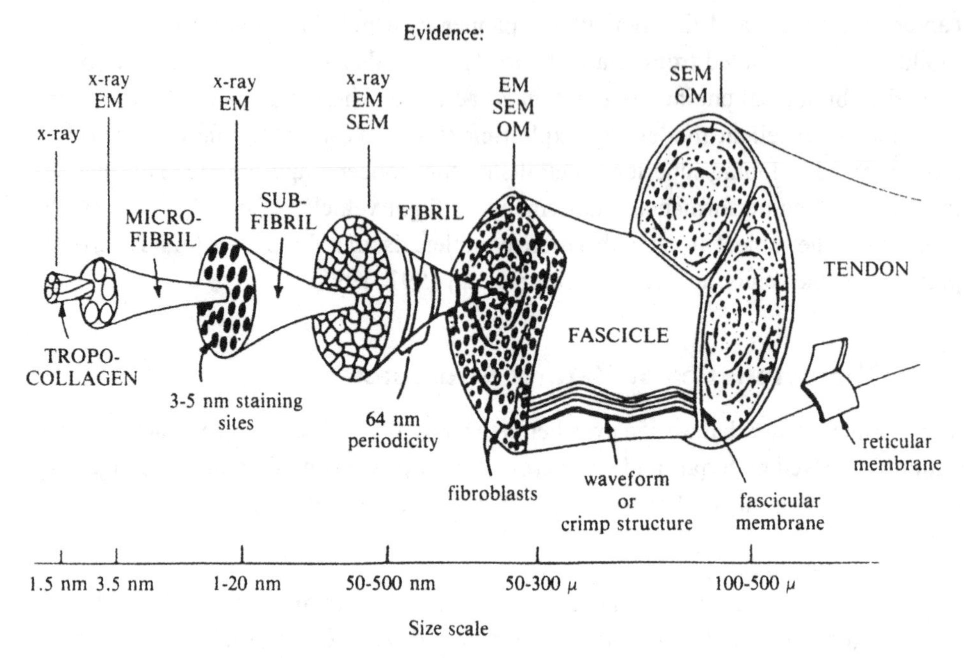

Figure 9.1. Tendon hierarchy (Kastelic and Baer, 1980).

complex structure, with six discrete levels of hierarchical organization, as shown in Fig. 9.1. The triple helix tropocollagen arrangement at the low end of the scale makes up the microfibrils, appropriately staggered and surrounded by a tetragonal lattice of subfibrils. These, in turn, are packed in and enveloped by additional fibrous structures as shown. This multilevel structural arrangement results in a tough structure with nonlinear and reversible properties.

Rope-like structures, which mimic to some degree the tendon architecture as shown in Fig. 9.2, are ancient man-made structural components that have withstood the tests of time in mining, marine, and elevator industries. Numerical results from a parametric study of a single strand (a central graphite fiber surrounded by six tungsten wires with varying values for the helical wrap angle) in tension indicate an increase in toughness with increased ductility of the element (Figs. 9.3 and 9.4). The parameter in these figures is the ratio of areas of the inner and outer fibers.

There are two important consequences of the subdivision and hierarchy of the elements discussed above. First, there is a loss of stiffness at the higher hierarchical levels. This reduction can be substantial. For example, in cellulose, Young's modulus of the crystallites has been estimated at 250 GPa, but the experimental value in well-aligned parallel systems of microfibrils is only of the order of 70 GPa (Jeronimidis, 1980a). A second consequence of the hierarchical arrangement appears to compensate for the loss of stiffness discussed above by an increase in toughness (i.e., in the interaction between the subelements at each structural level, inelastic deformations occur before fracture). Such

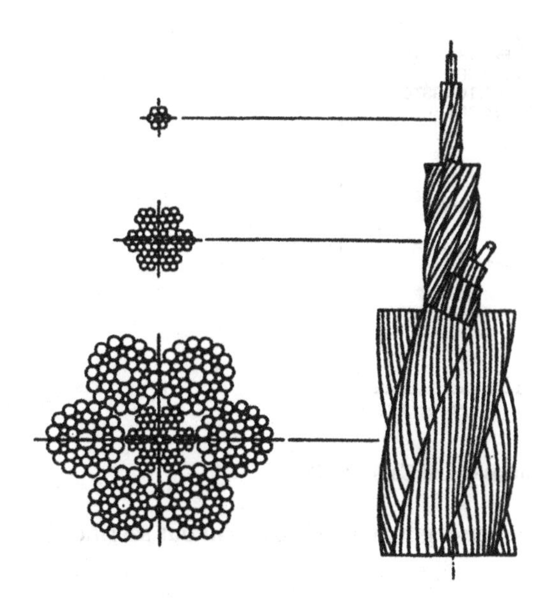

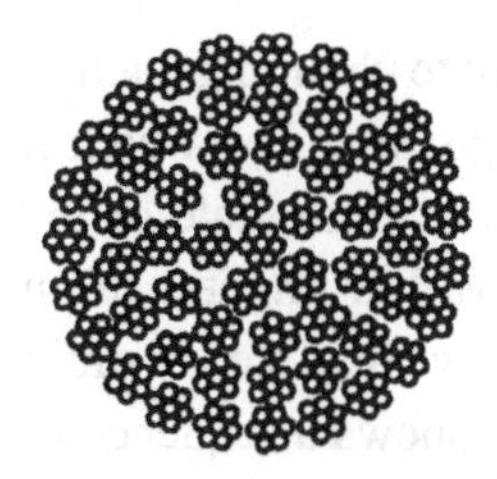

Figure 9.2. Rope architecture (Lee, 1991). (Reprinted, by permission, from Elsevier Science.)

deformations occurring at a very large number of interfaces contribute to energy absorption, leading to better structural integrity.

Thus, in natural systems, a judicious combination of elements, materials, and components of differing strengths in the same structure leads to acceptable and adequate hybrid systems whose properties are tailored for their specific usage. The variety of such designs is made possible by the hierarchical interaction between material and structure. This integrated approach seems to be one of the design rules and the key strategy with which nature achieves that elusive

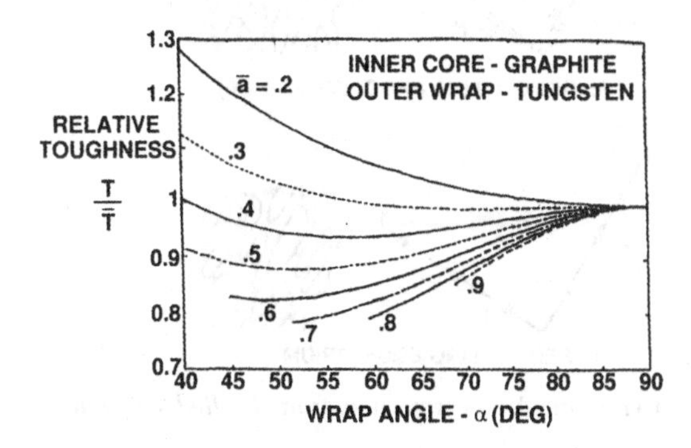

Figure 9.3. Rope architecture and toughness (Srinivasan and Cassenti, 1994).

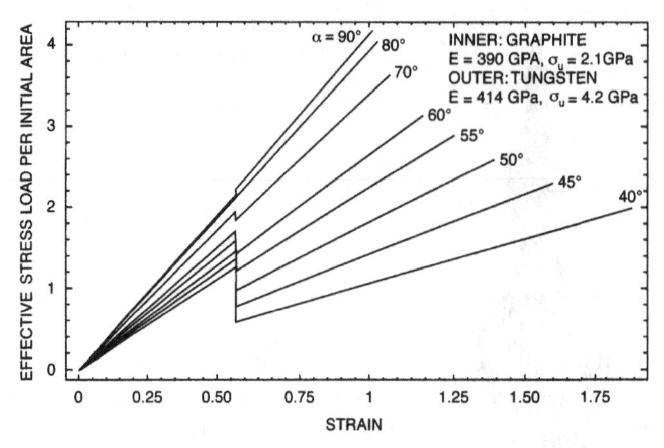

Figure 9.4. Rope architecture and ductility (Srinivasan and Cassenti, 1994).

but enviable objective of balancing strength and stiffness without sacrificing toughness.

3. *Adaptability*

   A good example of the ability of living organisms to adapt to changing loading conditions is bone. Living bone continually undergoes processes by which it remodels to accommodate the changes in its loading. The time scale of the process may be of the order of months but can begin within minutes of an impetus. Cyclic strain histories are found to be more effective than steady strain histories in inducing higher remodeling rates. Figure 9.5 shows a sequence of evolution in which structural adaptation takes place when the direction of load changes. A process that mimics this adaptation could be very useful in situations where sudden stresses are induced in certain sections of a structure, such as the wings of an aircraft during a high-speed maneuver. Obviously, any change in configuration or stiffness of the wing must occur very fast. One approach may be

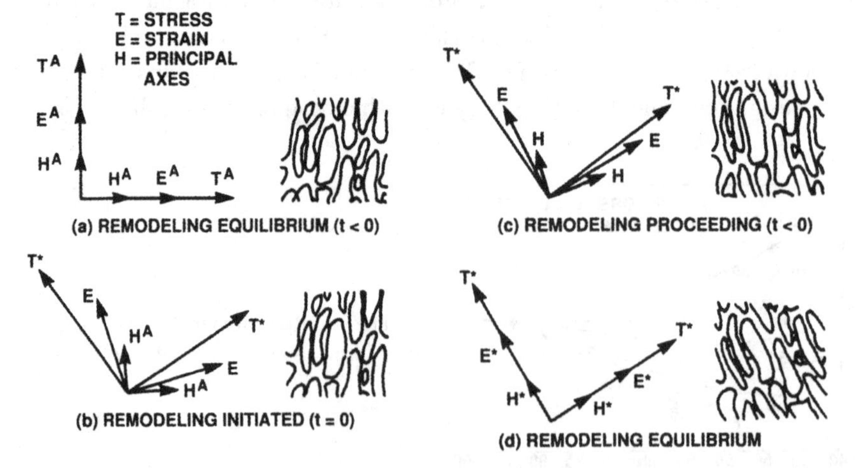

Figure 9.5. Adaptability of bone (Cowin, 1990). (Reprinted, by permission, from *Applied Mechanics Reviews*.)

utilization of a relatively rapid chemical process, such as crystallization. Clearly, rapid transfer of the materials to be crystallized to the site will be necessary. Circulatory systems in living organisms may provide some guidance in regard to this. Another interesting example of this characteristic is the realignment and reconfiguration of leaves on trees during a storm (Vogel, 1988). Unlike bone, where adaptive remodeling takes place in a slow continuous action, the structure of leaves allows quick deformation in response to the environment. Wind flow is made smoother as the leaves assume a conical shape to minimize the induced aerodynamic drag forces. As a result, the integrity of the whole tree is preserved.

The bull kelp that grows on the northwestern shores of the United States has a long (about 20 m), slender stipe, and it is subjected to a steady forced oscillation by wave action. The material of the stipe has a very low tensile strength (1 MPa), but it survives and thrives because the breaking strain in the material is not reached until it is stretched up to 30% to 45% of its resting length. The work required to break it was found to be 0.1 MJ m^{-2},which is an order of magnitude larger than that of stony corals (Wainwright, 1980). Stiffness at any cost is not a structural strategy often found in nature. The Eulerian buckling deformation of the hedgehog spines referred to earlier is another example of nature's strategy to survive through deforming.

Optimization of architectural, structural, and material arrangements in structural design is nature's way. In locusts, for example, the elastic energy needed to catapult themselves is mostly stored in the tendon of the femur of the hind leg. The tendon is well protected from surface flaws and imperfections. Therefore, the factor of safety can be low and was estimated to be about 1 (Bennet-Clark, 1975) in view of the low rate of loading, and low risk. In contrast to this, the sudden and unpredictable impact loads imposed on the horns of fighting billy goats require much more precaution in design. The core of bone is therefore sheathed by fibrous keratin. The latter has a special property: It becomes essentially insensitive to notches when it is damp. The goat is known to utilize this useful property by rubbing its horns in a dewy bush before it picks on its rival. A factor of safety of 10 is estimated (Kitchener, 1987) as nature's selection to meet the unexpected and sudden impact loads the horns may receive. Similar differences in properties between wet and dry conditions were also noted in nacre (Currey, 1980).

There are many other examples that illustrate how natural systems adapt to the environment. Bone, as we have seen, adapts slowly to a change in loading by changing its own mass and microstructure while maintaining its primary function. On the other hand, chameleon lizards are known to change their color almost instantly. Also, frogfishes can mimic particular objects in their environment by adapting their color to suit the objects such as rocks, sponges, sea weeds, and sea anemones (Pietsch and Grobecker, 1990). For aircraft applications, the time in which any reconfiguration, such as changes in attitude

> or color patterns for camouflage, takes place needs to be extremely short. Therefore, mimicking the process requires (a) a thorough understanding of the variables that influence, for example, bone remodeling and (b) identification of design concepts that permit the desired adaptation at the instant of a perceived threat. Therefore, although bone as a material may in itself be of little importance to engineers, unraveling the mechanisms that lead to its adaptability characteristics is of great interest.

Clearly, the characteristics discussed above are impressive, and the mechanisms contributing to those characteristics are quite complex. We may never be able to duplicate nature's sophistication. However, the strategy, the principles, and the optimization evident in these characteristics need to be studied, understood, and incorporated, to the extent possible, in the attempts to develop smart structures.

In the following section, four natural composites are described in some detail to illustrate some of the characteristics discussed above.

9.3 Biomimetic Structural Design

We begin by examining available test data for certain selected natural composites, specifically wood, cuticle, bone and antler, and mollusk shell. The materials have been arranged in this order to correspond both to their increasing modulus and to the change in chemistry associated with this increase in modulus. Wood and cuticle are considered as the plant and animal analogs, respectively, of fiber-reinforced, organic matrix composites. Bone and antler are a blend of fiber-reinforced, organic-matrix, and ceramic-matrix composites and may be considered as analogs of fiber-reinforced ceramers. Mollusk shells may be considered as analogs of ceramics although, as will be seen, they contain an extremely small but important component of organic material.

9.3.1 Fiber-Reinforced Organic-Matrix Natural Composites

9.3.1.1 Wood: A Plant Analog

Wood is a natural composite that exhibits a remarkable combination of strength, stiffness, and toughness. The unique hierarchical architecture with which the constituents of wood are arranged is the basis for achieving excellent properties. The manner in which wood is able to accomplish this is of special interest in this discussion. We begin with a brief discussion of key findings that led to a remarkable development of a patented new material inspired and guided by a thorough understanding of the interplay between morphology and structural performance of different types of wood.

Cellulose, which is a high molecular weight polysaccharide, is the main constituent of wood and is directly responsible for stiffness and strength. Cellulose is made by stringing glucose molecules together. It is important to note that both crystalline and amorphous regions coexist in the arrangement of cellulose molecules. In addition, there are varying quantities of low molecular weight sugars and a binding matrix known as lignin. The stiffness and strength of cellulose itself are high; theoretical values for Young's modulus and tensile strength are quoted to be 250 GPa and 25 GPa, respectively. Lignin is a heavily cross-linked phenolic resin and is very brittle (Jeronimidis, 1980b).

Wood has a cellular composite structure in which it is possible to identify four levels of organization: molecular, fibrillar (ultrastructural), cellular (microscopic), and macroscopic (Jeronimidis, 1980b). A wood cell is essentially a hollow tube of about 30 μm diameter with a multilayered laminated wall. The innermost layer is the thickest (close to 80% of the total cell wall area) and is the principal load-bearing component. It contains the cellulose fibers in the form of microfibrils of about 10 to 20 nm in diameter. In most cases, the fibrils do not lie parallel to the axis of the cell but form a steep helix (Fig. 9.6) at an angle between 5°and 20° to the grain direction. Also, in any one tree these fibrils are wound in the same way. The outer layers are much thinner with a cross-helical fiber arrangement (fiber angle 40° to 60°) except in the outermost layer, where it is random. There is some evidence to suggest that these outermost layers help in stabilizing the inner thick layer against buckling in compression (Jeronimidis, 1991). The cells are parallel to the grain direction and are bonded to each other by a phenolic matrix, which is also present in the cell walls in various amounts.

Wood is stronger in tension than in compression and has a breaking stress of about 100 to 140 MPa. The density of wood is around 1/20 that of steel and therefore the relative tensile strength of wood is equivalent to a steel having a tensile strength of about 3 GPa. Even the "high tensile steels" do not have tensile strengths much above 1.75 GPa, and they are relatively brittle (Gordon, 1988). The compressive strength of

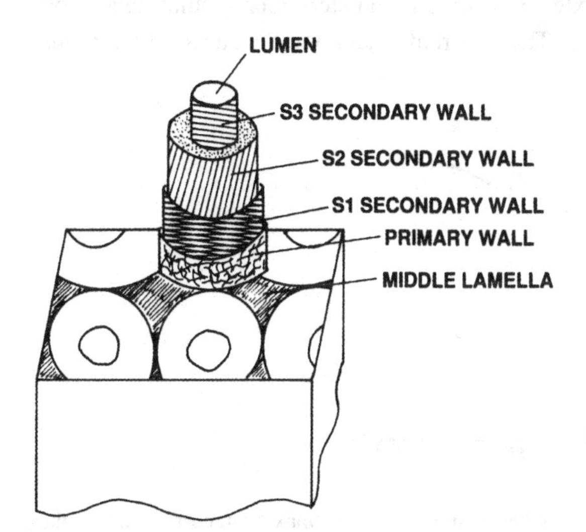

Figure 9.6. Simplified model of a wood tracheid (Jeronimidis, 1980b). (Courtesy of the Royal Society, UK.)

wood is about a third of its tensile strength. The difference is attributed to local buckling of the cell walls under compression followed by a macroscopic crease.

Among the important parameters that influence mechanical properties of wood are: (a) properties of the constituents, (b) microfibrillar angles, (c) volume fraction, (d) direction of applied load (i.e., longitudinal, radial, or tangential; longitudinal values of E can be 20 to 60 times greater than those in the radial and tangential directions), and (e) the specific laminated unsymmetrical structure of the cell wall in each cell (Jeronimidis, 1991). Being an anisotropic material, wood has fracture properties strongly dependent upon the direction of loading. Two directions are evidently of interest: one along the grain and the other across it. It is common knowledge that wood is weak along the grain. The splitting of wood along the grain is a brittle phenomenon. Fracture energies R_f of 0.1 to 0.2 kJ m^{-2} have been measured. Such a low value of work of fracture means that the critical crack lengths are of the order of a few millimeters only. However, fracture across the grain requires energies as high as 12 kJ m^{-2}. For example, specific work of fracture (R_f/ρ where ρ is specific gravity) of Sitka Spruce ($r = 0.4$) is 30 kJ m^{-2} and compares very well indeed with the lower end of the range of the corresponding values for steel, which varies between 12.8 kJ m^{-2} and 128 kJ m^{-2} (Jeronimidis, 1980a,b).

Initial studies to understand the mechanisms of failure assumed fiber pull-out to be the principal mechanism. However, optical and electron-microscopic examination of fractured surfaces did not show any evidence of wood cell pull-out. Therefore, a different energy absorption mechanism in wood was postulated and eventually verified. The new mechanism took into consideration energy losses that could be attributed to work done in the process of buckling of some cells. It is known that a hollow tube helically wound with elastic fibers is unstable and when pulled the fibers in the cell walls tend to straighten and in the process collapse inward, leading to a sudden drop in load (buckling) followed by considerable postbuckling elongation at a load lower than the maximum. A typical load-extension curve for a macrofiber simulating a wood cell with $\varphi = 15°$ is shown in Fig. 9.7. The apparent "yield point" in a wood fiber thus

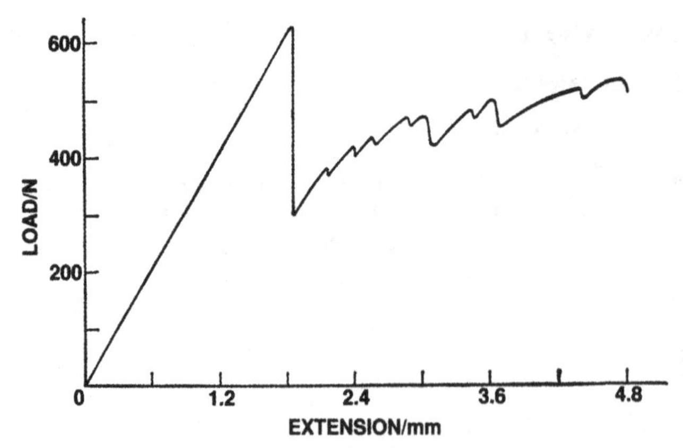

Figure 9.7. Load-extension curve for a macrofiber with $\theta = 15°$. The maximum corresponds to the onset of buckling (Gordon and Jeronimidis, 1980a). (Courtesy of the Royal Society, UK.)

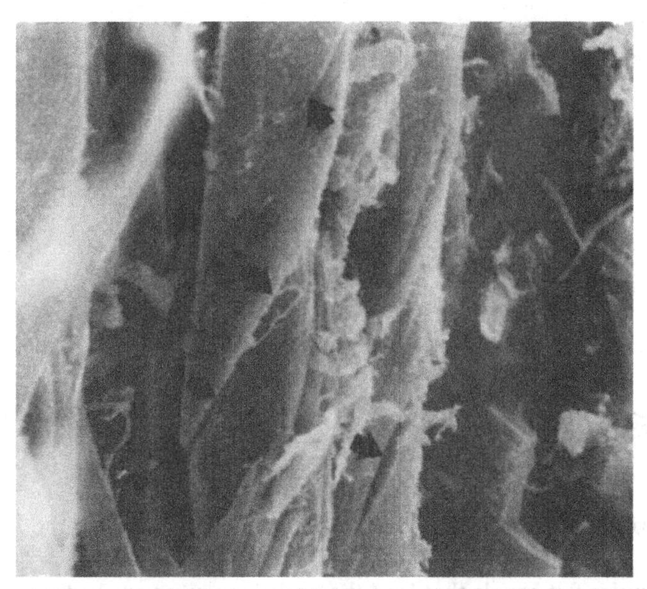

Figure 9.8. An SEM image of spruce wood (Jeronimidis, 1980a). (Courtesy of the Royal Society, UK.)

corresponds to the sudden buckling of the cell walls under tension. After buckling, such a tube will not, as a rule, go back to its original form and so may show a very large absorption of energy. Only about 10% of the fibers are observed to undergo this phenomenon. Such a tensile failure, shows both a folding inward as well as debonding, is clearly illustrated in Fig. 9.8. In wood the energy absorbed per cell can be estimated to be of the order of 2×10^{-4} J. As the number of cells per square meter in wood is about 10^9, the fracture energy per unit area is approximately $200\,\text{kJ}\,\text{m}^{-2}$. This is an order of magnitude higher than the experimental values but has been obtained on the basis that all the cells in the cross section undergo the same phenomenon. In fact, if only 10% of the cells are deforming in the manner stated above, then the calculated and measured values do indeed agree quite well. In any case, the important point to note here is that in wood the energy absorbing mechanism depends on the detailed arrangement of the cellulose fibers in the cell walls. These studies to understand the sources of energy loss to account for the measured higher fracture energy have led not only to a more sound model but, more importantly, to innovative ways of constructing strong and tough composites.

This understanding led to a promising man-made counterpart composite of fiber-reinforced plastic. Such a composite material/structure was patented (Chaplin, Gordon, and Jeronimidis, 1983) and was based on the concept of a glass fiber-reinforced plastic composite of helically wound man-made fibrous elements, such as glass or carbon hollow tubes. In this concept, the fibers within the tube walls and the tubes themselves are bonded together by a polymeric matrix with the tube cores remaining empty. With an optimum helix angle of 15° for the fiber orientation, the resulting composite construction was found to achieve a high value for specific work of fracture

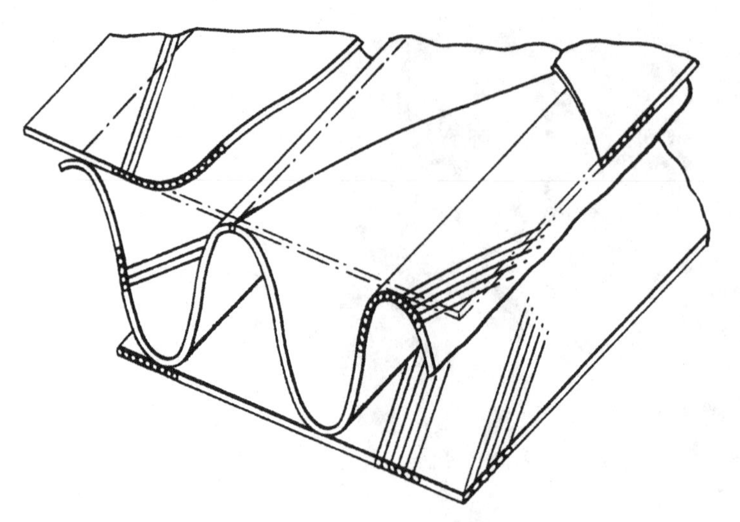

Figure 9.9. Wood type composite concept (Chaplin et al., 1983).

of 400 kJ m^{-2}. Somewhat simpler flat panels that lend themselves to mass production, shown in Fig. 9.9, were developed and tested in strips of 2 in. × 10 in. With the same optimum angle, specific work of fracture of up to 109 kJ m^{-2} was measured.

9.3.1.2 Insect Cuticle: An Animal Analog

The two primary components of insect cuticle are (a) chitin fibers and (b) a proteinaceous matrix. The fibers are high aspect ratio rods of about 3 nm in diameter. The matrix is formed primarily from various proteins and other organic and inorganic materials. The stiffnesses of chitin fibrils and the matrix protein are reported to be between 70–90 GPa and about 120 MPa respectively (Vincent, 1991).

The cuticle is divided into two primary sections, the epicuticle and the procuticle. The epicuticle layer is a protective coating whose thickness may vary from 0.1 to 10 μm and is the essential protective barrier. The structural part that provides shape and mechanical stability is the procuticle. The thickness of the procuticle may vary from 10 to 200 μm and it is made up of a series of thin composite lamellae containing chitin fibers, which serve as reinforcements. Within each layer the chitin fibrils are about 6 nm apart and oriented in the same direction. Whether the direction changes from layer to layer will depend upon the circumstances. For example, if the forces normally experienced by a part are along a single direction (as in the locust tendon), then the direction of chitin fibers is clearly preferred in that direction. However the orientation of the fibrils usually changes from layer to layer. For example, in the water bug *Hydrocirus*, the individual fibrils are about 4.5 nm in diameter and positioned about 6.5 nm apart. This is also the thickness of the layer. The preferred orientation of neighboring sheets varies about 7° or 8° and always in the same direction (clockwise or counter-clockwise) throughout the cuticle. As a result, the preferred direction changes by 180° in about 25 layers, a thickness of 160 nm. Therefore, in a cuticle more than a few microns thick, the fibers will be pointing in many directions in the plane of the

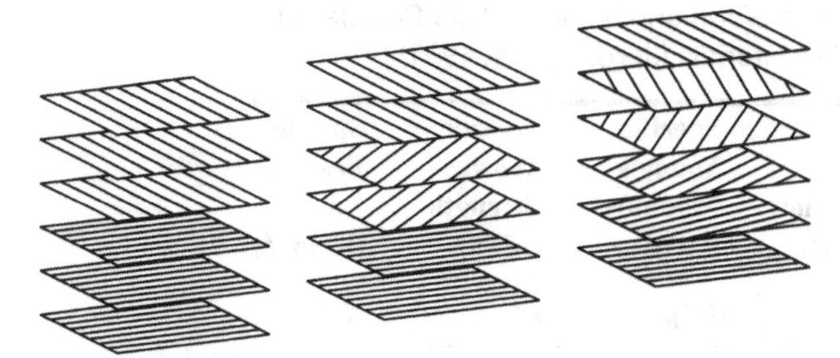

Figure 9.10. A cross section of a generic insect cuticle (after Vincent, 1991).

cuticle, as shown in Fig. 9.10. It should be noted that the orientation of chitin around holes (such as dermal glands or pore canals) and other such discontinuities varies in a rational manner to carry the stress around the obstacle to minimize the effect of stress concentration (Fig. 9.11). The intricacies of the arrangement of fibers in general and around stress risers in particular offer important guidelines in the design and fabrication of fiber-reinforced composites.

Insect cuticle is an extraordinary natural composite assembled at the nanometer level. The sophistication and control evident in its fabrication are simply remarkable. One marvels at the processes that allow a range of seven orders of magnitude in Young's modulus and extremely intricate differences in architecture over the order of 1 mm or less, using essentially the same materials (chitin and protein; Vincent, 1991).

In Table 9.1, a concise summary of the properties of wood and insect cuticle is tabulated for easy reference. Corresponding properties of mild steel and of glass fiber-reinforced and graphite fiber-reinforced composites are included in order to highlight the comparison with natural materials. Specific properties (property/specific gravity) for Young's modulus, tensile strength, and fracture energy are shown (Jeronimidis, 1991).

9.3.2 Fiber-Reinforced Natural Ceramers: Bone and Antler

Bone typically forms the skeleton that provides support and protection for vertebrates (fishes, reptiles, birds, and mammals). As a biological material system, it is of great interest because bones exhibit specific mechanical properties comparable with those of

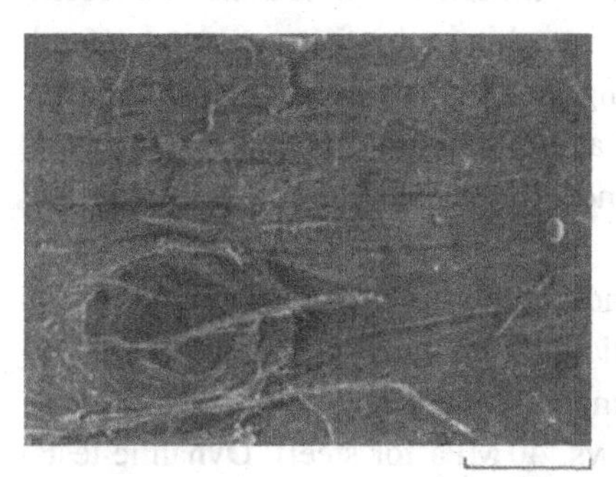

Figure 9.11. A high-magnification view of a pore canal (Gunderson and Schiavone, 1987). (Reprinted, by permission, from the *Journal of Materials Science*. Copyright Kluwer Academic Publishers.)

Table 9.1. Mechanical Properties of Wood and Insect Cuticle Compared with Those of Certain Man-Made Materials (Jeronimidis, 1991)

Material	Specific Gravity ρ	Specific Young's Modulus (GPa) E/ρ	Specific Tensile Strength (MPa) s/ρ	Specific Fracture Energy (kJ m^{-2}) R_f/ρ
Mild Steel	7.9	25	38–51	13–128
Glass-Fiber Composites	2.0	17–22	500–600	15–25
Graphite-Fiber Composites	1.5	87–135	670–800	7–20
Wood	0.4–0.6	25–30	170–220	13–50
Cuticle	1.2	10^{-6}–14	60–125	0.2–1.8

metals. However, bone is far more complicated than metals and possesses remarkable abilities of structural adaptation to external loading. Bone, just like all biological material systems, is designed for specific functions and, therefore, its mechanical properties vary not only from animal to animal but also from one location to another in the same animal.

The basic constituents of bone are a protein known as collagen and an inorganic mineral phase called hydroxyapatite, consisting primarily of calcium phosphate crystals. Collagen is tough and has excellent tensile properties but is also flexible (only 1% as stiff as the mineral with about 1 MPa stress at a strain of 10%) and therefore unsuitable to carry compressive or bending loads. Therefore, early on in the growth process, thin platelets of ceramic are added into specific holes within the collagen. Thus, even at the microstructural level the material is nonhomogeneous. Hydroxyapatite is twice as strong and much stiffer ($E = 130$ MPa with an ultimate strain of 10%) than collagen, and thus the resulting composite has around 67% of the inorganic component in most adult bones in the form of thin plates (Gordon, 1988).

The complex architecture of bone is shown schematically in Fig. 9.12. The structural unit that corresponds to cells in wood is the osteon, which, as a hollow tube, serves not only as a load-bearing component but also as a conveyor of blood. The walls of the osteon are constructed much like wood cells with a helical arrangement of thin fibers of collagen alternating with parallel fibers of hydroxyapatite. These fibers "are, in effect, equivalent to whisker crystals" (Gordon, 1988). It is important to note that more recent thinking in the research community is that the mineral is in the form of platelets and not fibers (Weiner and Traub, 1991). The resulting material is a short column capable of transmitting compressive loads effectively to avoid buckling. However, in view of the adequate protective covering bone has around it, it is not designed primarily for toughness.

Osteons are typically 200 μm in diameter and about 1 to 2 cm long. The orientation and density of the mineralized fibers are dictated by the stress that needs to be sustained at a given location. The resulting composite has nearly twice the specific compressive strength of steel (75 MPa for bone vs. 40 MPa for steel). Dynamic tests

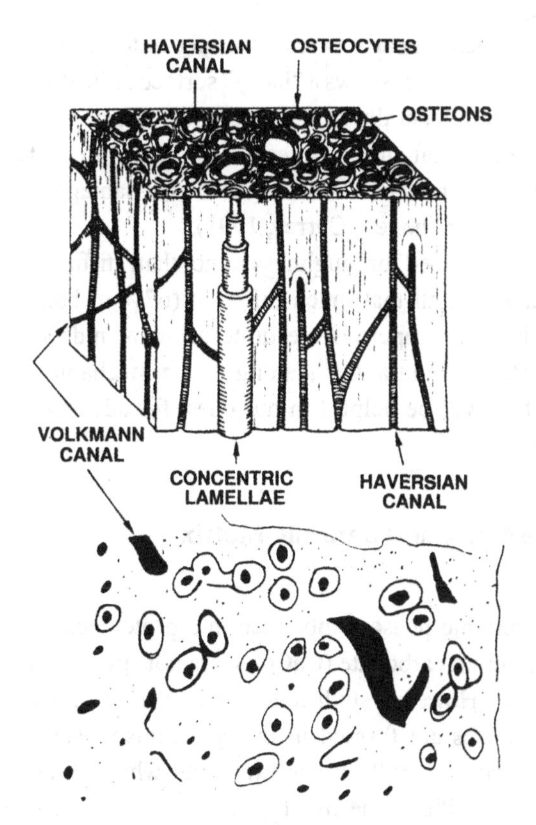

Figure 9.12. Schematic of bone microstructure (Sadananda, 1991). (Reprinted, by permission, from the *Journal of Materials Research*, Materials Research Society.)

with the Hopkinson bar technique conducted on human cortical bone show that the dynamic Young's modulus was 23% higher (19.9 GPa) than the corresponding static value (16.2 GPa). The value of Poisson's ratio remained the same at an average value of 0.36. Bone displays a viscoelastic behavior and tests show that any stress wave will attenuate as it propagates along the material resulting in near total absorption of energy in about 500 μs (Katsamanis et al., 1990). However, its specific work of fracture does not compare very well with either wood or steel (1 000 Jm^{-2} compared to 37 000 Jm^{-2} for pine and 10 000 to over 100 000 Jm^{-2} for mild steel). It may be recalled that the high work of fracture measured in wood was attributed to the ability of wood cells to buckle under a tensile load. In the case of bone, the osteon walls are much thicker and therefore resist buckling. The mechanism of fracture is essentially one of energy dissipation through relative motion at the weak interfaces between the osteons (Gordon, 1988).

For stresses of the order of 100 MPa, the critical Griffith crack length for bone is around 3 cm (compared to 50 cm for wood and 1–2 m for steel) and is the same for both small and large animals. Therefore, small animals such as mice, cats, and dogs with bones less than a couple of centimeters thick do not generally experience bone fracture, unlike horses and humans, who are more susceptible to bone fracture (Gordon, 1988).

The strength and toughness properties of antler are much better than those of bone. Specific work of fracture equal to 7 500 Jm^{-2} and specific compressive strength of

120 MPa have been reported. Similarly, antler is nearly twice as strong in tension as bone. An examination of the fracture surface of antler shows a "hairy" surface indicative of extensive fiber pull-out. This is attributed generally to detailed and sophisticated control over the fine fiber structure of the material and the interface adhesion. Recent creep rupture test results indicate that the "rate at which antler accumulates damage at any strain is far less than that shown by standard bone" (Currey, 1991).

Unlike bone, the structure of antler is optimized for toughness rather than stiffness. The mineral content in bone and antler is not all that much different (67% in bone and 55% in antler), and yet their mechanical properties are quite different indeed. The morphology and the architecture of bone and antler leading to the mechanical properties discussed above offer clues that will be helpful in our quest for advanced material/structure systems.

9.3.3 Fiber-Reinforced Organic-Matrix and Ceramic-Matrix Composites: Mollusks

Mollusk shells consist of one or more ceramic phases embedded in a proteinaceous matrix. These ceramic phases, such as calcium carbonate ($CaCO_3$), are not suitable as structural materials because they are brittle. However, studies of mollusks have shown that remarkably strong and tough composites are formed in living organisms starting with weak constituents. Five main types of mollusk shell material which differ in structure and composition have been identified. The five types, prismatic (polygonal columns), nacreous (flat tablets), crossed lamellar (plywood-like), foliated (long thin crystals in overlapping layers), and homogeneous (fine-scale rubble), can occur alongside each other.

The abalone shell has a "brick and mortar" structure with $CaCO_3$ bricks occupying 95% of its volume with the remaining 5% of the microstructure being organic matrix serving as the mortar. Two layers are present in the microstructure, an outer prismatic layer, which is essentially calcite (rhombohedral form of $CaCO_3$), and an inner nacreous layer of aragonite. It is this nacreous layer that is found to display a good set of mechanical properties and was therefore the subject of study by Sarikaya et al. (1990). Estimates of tensile strength and toughness of the inorganic component vary but values of 100–150 MPa for strength and R_f as high as 0.01 kJ m^{-2} are not uncommon. The organic component is estimated to be superplastic and about 20–30 nm thick (Sarikaya et al., 1990).

There is a tendency for cracks in these shells to be temporarily arrested at the boundaries between the layers. This is principally due to the ingenious way in which the easy and difficult directions of fracture, depicted in Fig. 9.13, are integrated into the structure. In the easy direction, the cracks need only to pry open two planes. In the difficult direction, cracks have to pass through the lamellae. As neighboring laths are oriented at different angles, the crack front gets continuously redirected. Dictated by the microstructure, the crack path thus becomes tortuous, and toughness is increased by this crack branching. In the tests on *Strombus*, laths were observed protruding from

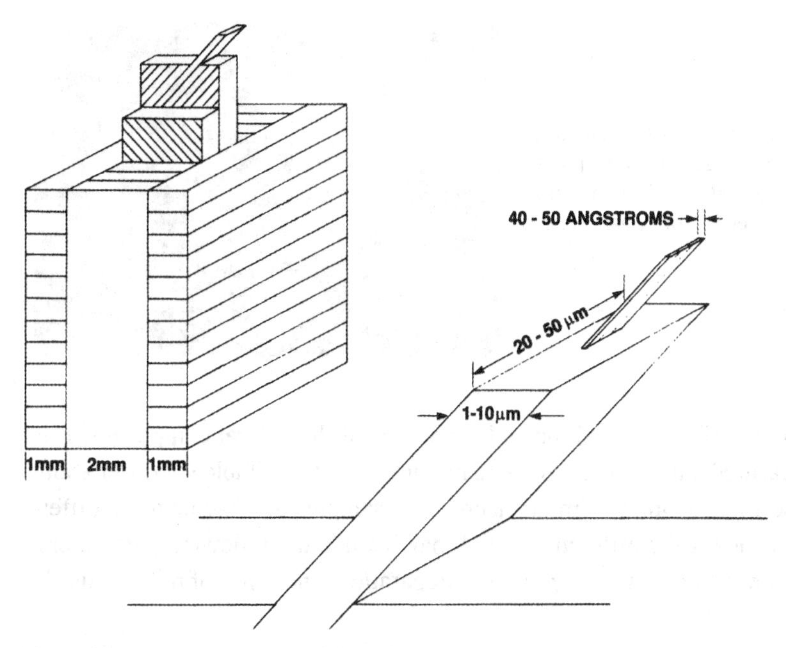

Figure 9.13. Structure of crossed-lamellar shell (Laraia and Heuer, 1991).

the cracked surface, indicating a fiber pull-out type mechanism (Fig. 9.14). To sum up, toughening mechanisms that have been observed include plate sliding and pull-out, microcracking along interlamellar boundaries, stretching of the organic microfibrils resulting in crack bridging, and delamination cracks just ahead of the main crack tip (Figs. 9.15 and 9.16). The microstructure design leads to the variety of toughening mechanisms acting in concert, resulting in a tough material. The high degree of order in the morphology, orientation, and spatial distribution of the ceramic phase, and interfacial

Figure 9.14. Lath "pull-out" in fractured *Strombus* (Laraia and Heuer, 1989; Reprinted with permission of The American Ceramic Society, Post Office Box 6136, Westerville, Ohio 43086–6136. Copyright 1989 by The American Ceramic Society. All rights reserved.)

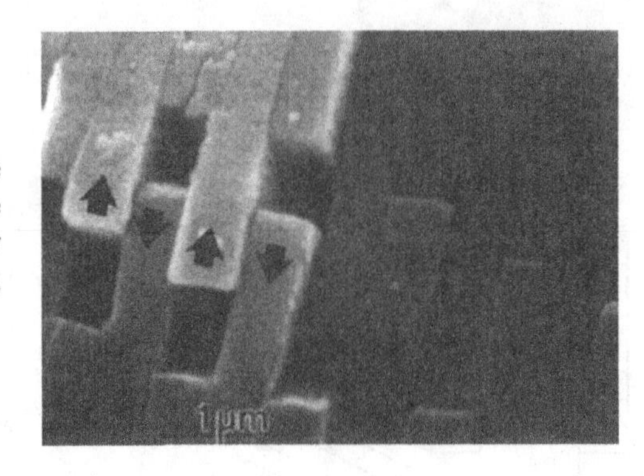

Figure 9.15. SEM micrograph of an abalone shell test sample revealing sliding in the plane of the layers (Sarikaya et al., 1990). (Courtesy of M. Sarikaya, University of Washington, Seattle.)

characteristics all contribute toward the end properties of the natural composites. It is not clear why measured values of stress intensity factor shown in Table 9.2 differ about 100% from those for essentially similar nacre. The samples are undoubtedly different, but other reasons for the difference noted may include the following parameters: span/depth ratio, level of hydration, specimen preparation, and types of notches used.

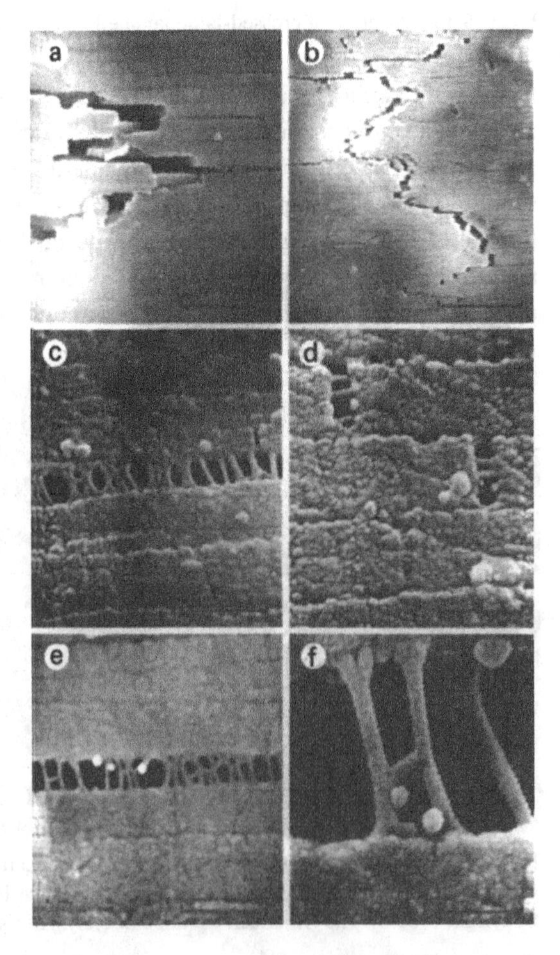

Figure 9.16. Toughness mechanisms in nacre: (a) Delamination cracks extend laterally from the main crack. (b) Delamination cracks just ahead of the main crack tip. (c) Fibrils of organic matrix bridging the platelet lamellae across a delamination crack. (d) Fibrils of organic matrix extending between the ends of the platelets. (e) Fibrils of organic matrix. (f) Fibrils of organic matrix bridging a delamination crack. (Courtesy the Royal Society, Jackson et al., 1988.)

Table 9.2. Mechanical Properties of Mollusk Shells (Jackson, Vincent, and Turner, 1990)

Shell	Tensile Strength (MPa)	Young's Modulus (GPa)	Mean Fracture Energy ($J\,m^{-2}$)	Stress Intensity Factor
Strombus	100	60	280[c]	4[c]
Abalone	180	60[c]	200[c]	7 ± 3
		70	1600	
Pinctada	275[a]	60[a]	180[b]	2.4[a]
	224	73	1500	5.9

[a] The range of values depends upon span/depth ratio, hydration, and direction of crack propagation, i.e., across or along platelets. For a thorough list see Tables III, IV, and V (Jackson, Vincent, and Turner, 1990) and Table 4 (Jackson et al., 1988).
[b] Lower values are for crack growth along the platelets.
[c] Estimates only.

Some interesting and possibly significant observations can be made by examining data reported in Table 9.2 (Jackson, Vincent, and Turner, 1990). Comparing work of fracture for $CaCO_3$ (assumed to be 0.5 Jm^{-2}) and that for *Pinctada* (about 1500 Jm^{-2}), a three thousand-fold increase becomes possible because of the reasons stated above. Comparing nacre with mild steel, for example, the data shows E/ρ for both the materials to be essentially the same: about 25 GPa. Similarly, the strength parameter s/ρ for nacre is about 90 MPa, nearly twice that for steel, which is 50 MPa. These numbers are impressive considering the weak constituents with which the shells were assembled. Inspiration derived from nacre's construction has led to innovative processing concepts for man-made materials, as can be seen by the results shown in Fig. 9.17. For the boron-carbide-aluminum (B_4C-Al) cermet with a 70% boron carbide and 30% aluminum composition, toughness measurements of about 0.7 $kJ\,m^{-2}$, (K_I, about 15 $MPa\,m^{1/2}$) have been obtained with a corresponding fracture strength of about 1 GPa (Fig. 9.17). The trend is very promising and illustrates the extent of possibilities for man-made composites guided by natural systems.

It is interesting and informative to make one more comparison of the properties of mollusks with those of some of the more advanced man-made ceramic composites. The latter have led to a remarkable decade of development, culminating in materials having toughness of the order of 20 $MPa\,m^{1/2}$ (Evans, 1990). The mechanisms they exploit to attain high toughness values include transformation, microcracking, metal dispersion, whiskers/platelets, and fibers, leading to values for R_f in excess of 6 $kJ\,m^{-2}$ ($K_I > 30\,MPa\,m^{1/2}$). As Evans notes, the numbers are impressive especially when one appreciates that conventional ceramics have a toughness of only 1 to 3 $MPa\,m^{1/2}$. A cautious note is struck that "these highest levels cannot usually be used to effectively enhance strength and reliability." Table 9.3 provides a concise summary of properties of mollusk shells, bone/antler, and certain selected engineering ceramics with reference to corresponding properties of steel (Jeronimidis, 1991).

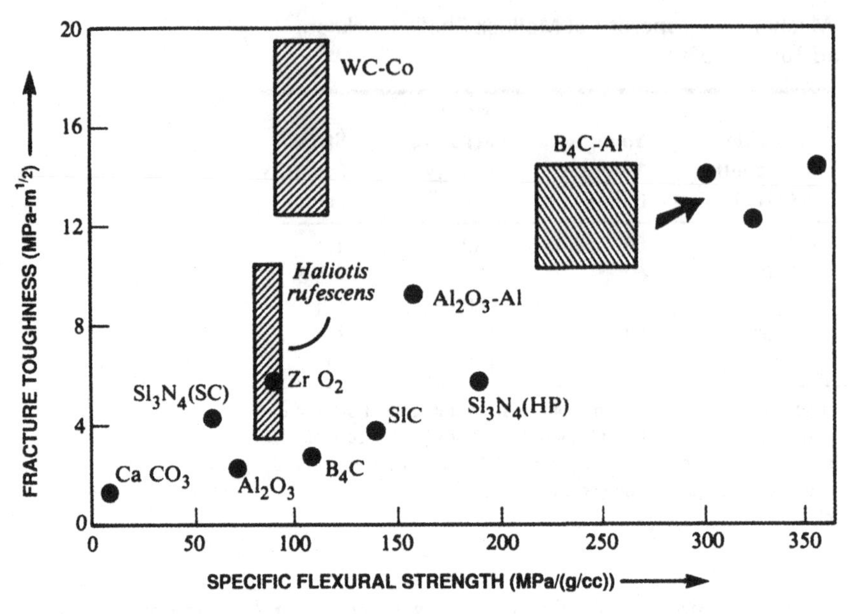

Figure 9.17. Toughness of abalone shell (Sarikaya et al., 1990). (Courtesy of M. Sarikaya, University of Washington, Seattle.)

9.4 Biomimetic Sensing

Before closing this chapter, another aspect of smart structures, sensing parameters, is studied with three analogs from natural systems, cochlea, bats, and arachnids, in order to understand the principles used by natural structures to sense precisely parameters of engineering interest.

Man-made devices that sense engineering variables such as shapes, distances, sound, strain, and light become integral parts of major structural systems such as submarines, aircraft, engines, and photographic equipment. There are literally thousands of biological systems in which sensing of these variables takes place, followed by processing and actuation.

Table 9.3. Mechanical Properties of Natural Materials Compared with Certain Man-Made Materials (Jeronimidis, 1991)

Material	Specific Gravity ρ	Specific Young's Modulus (GPa) E/ρ	Specific Tensile Strength (MPa) s/ρ	Specific Fracture Energy (kJ m^{-2}) R_f/ρ
Mild Steel	7.9	25	38–51	13–128
Engineering Ceramics	2.3–3.0	50–170	67–220	0.0003–0.04
"Exemplary" Ceramics[a]	2.87–3.82	75–130	Unavailable	2–6
Bone/Antler	2.0	7.5–10	60–80	2–7
Mollusk Shells	2.8	21–25	36–96	0.07–0.6

[a] (Evans, 1991).

9.4.1 Cochlea

The first example pertains to a study of the electro-elastic behavior of auditory receptor cells by Steele (1992). The cochlea of the inner ear is a transducer: a mechanical input to the cochlea from a sound source produces an electrical output. In the hearing process, certain receptor cells in the cochlea undergo substantial elongation with change in the electrical field, leading to an enhancement of low-amplitude sound. Such displacement was observed to be five orders of magnitude larger than that which can be measured by a piezoelectric crystal of the same size and for the same strength of the electrical signal. The superior performance of the receptor cells is attributed to the "unique microstructure design of the wall of the cell." Steele concludes that "in comparison with man-made devices, unusual principles of design are being used to obtain unusual effects" and predicts that "even a partial understanding of the principles may motivate new thought on the design of devices."

9.4.2 Bats

If dragonflies and birds provide analogs of flight system, bats offer a superb analog for precision sensing. Until recently, a common misconception was that bats' use of sound pulses to navigate and locate prey is a "crude system, acoustic equivalent of feeling one's way in the dark with a cane" (Attenborough, 1990). But biosonar "design" has since been shown to be anything but crude: an echolocating bat can pursue and capture a fleeing moth with a facility and success rate that would be the envy of any military aircraft designer. In addition to providing information about how far away a target is, bat sonar can relay some remarkable details. Doppler shifts – changes in the frequency of the echo, relative to the original signal – convey information not only about the relative velocity of a flying insect, but also about its wingbeat. The amplitude of the echo, combined with the delay, indicates the size of the target. The amplitudes of the component frequencies correspond to the size of various features of the target. The differences between the ears in intensity and arrival time of sound give the azimuth of the target, whereas the interference pattern of sound waves reflected within the structure of the outer ear gives the elevation.

David Attenborough, in his famous book, *Trials of Life* (Attenborough, 1990), tells a fascinating story about bats. It appears that in the wilds pregnant bats seek nursery accommodations in caves that are warm and humid and with a temperature of about 100°F that is essentially constant through night and day. One such site is Bracken Cave in Texas, which is said to hold over 20 million bats. The babies are parked in a special nursery area in the back of the cave. They are packed at about 1000/yd^2. Mother bats fly out to feed and return to feed their babies. Not too long ago, it was believed that mother bats had no preference and fed whoever got to them first; a socialism type theory, if you will, in the bat world, where everyone shared without any preference. As it turns out, not only does the mother bat care, but she uses high technology to locate her baby and won't feed any other baby. Thus, when she returns, she has an engineering problem, which is to locate her baby, in what must surely be a chaotic, confusing, raucous racket. She alights within a few feet of where she last left her baby. It is unlikely to be in the same position as the babies will have, it seems, moved about 18 in.

> As she lands, she calls for several seconds and her baby answers. It is difficult to believe that, among the tumult in the cave, either mother or baby would be able to recognize one another's voices, but bats are famous for their skill in disentangling the echoes of their high frequency squeaks and using them as a way of navigating. Individual calls vary greatly. Slowed down to the frequencies that suit our ears one can hear that they differ in volume, length, pitch and frequency, and include squeals, yelps, grunts and trills. (Attenborough, 1990)

One can imagine the signal-to-noise ratio, the filtering, the processing, the transforms that take place in seconds. It is truly a remarkable piece of technology, all taking place inside a brain no larger than a pearl.

9.4.3 Arachnids

The third and final example in this category is the slit sense organ in arachnids, studied by Barth (1978). These sense organs are not unlike strain gages, and are located in "holes" in the cuticle. Their shapes vary from circular to elliptic to elongated slits covered by membrane to which sensory cells attach. Loads on the exoskeleton due to its own weight, muscular activity, and variations in the environment are sensed by these "gages" through the process of deformation of the holes, leading to deformation of the membrane and finally to the cells where a nervous response is triggered. The elongated shape of the slit allows a preferential directionality of deformability in which it becomes most sensitive to loads perpendicular to its long axis. The slits are located roughly perpendicular to the lines of principal stresses. The "design" is such that the slit lengths vary from about 5 μm to about 200 μm, leading to a broad spectrum of sensitivities. The membrane cover is extremely thin, about 0.25 μm, with a slight curvature, another example of an "unusual design principle" to accomplish "unusual effects."

9.5 Challenges and Opportunities

Biologically evolved structures and materials promise to provide insights and ideas that will permit us to achieve the manifold improvements necessary in realizing the envisioned advanced systems. There are many challenges and opportunities in such an endeavor. For example, military aircraft of the next century are expected to operate at sustained supersonic regimes, at altitudes of 100,000 ft or more and in a wider combat arena than at present. The goals set for powerplants are equally ambitious. A doubling of propulsion capability (thrust/weight) in the early years of the next decade may not be unrealistic. Engines with that capability will require materials that can perform at temperatures up to 4,000 °F with little or no cooling requirement. Current concepts for the National Aerospace Plane demand similar high temperature materials.

Revolutionary changes in material and structural design are needed if we are to meet the technological challenges that loom ahead in developing emerging aerospace systems. Biological structures and materials promise to provide insights and ideas that will permit us to achieve the improvements necessary in realizing the envisioned advanced systems. Some of these challenges are discussed below and the opportunities such challenges open up should be self-evident.

Research efforts should be aimed at not only understanding the hierarchical material construction but also unraveling the mysteries of the process by which it is achieved in natural systems. The latter is much more complex and possibly involves neural aspects of natural systems from birth to full development. What mechanisms, for example, control the deposition of zinc at the cutting edge of locust teeth? What sophisticated and precise control mechanism do we have in our jaws as we bite into a nut? What processes are used in living organisms to control the microstructure of materials? What processes are used to obtain strong and tough material/structural systems starting from relatively weak constituents? The unlocking of these secrets may well hold the key to truly innovative manufacturing and material systems development. Indeed, understanding the natural phenomenon of self-assembly may prove critical for mimicking the evolution of natural structures with any reasonable degree of sophistication.

In order to understand how material in any living system works, we need to conduct tests as well as model the system analytically. In experimenting with natural systems, the usual problems of specimen size, type, preparation, loading, fixtures, types of instrumentation, level of hydration, and so on assume unusual proportions. An understanding of the failure mechanisms of natural systems is crucial to structural design of advanced systems if we are to derive the same advantages in regard to durability that nature appears to have built into living systems. As observed in the specific natural composites studied above, the process of failure reveals an elaborate scheme to resist and dissipate the progress of fracture. Analytical modeling of failure mechanisms may require a knowledge of the lowest level in the hierarchical arrangement that needs to be represented. Are the mechanical properties that influence durability governed by processes at the molecular level? Should the finite element of the future be an atom? Or a grain consisting of thousands of atoms? When you examine the architecture of tendon, for example, you wonder if there are similarity laws among the various levels from the nanoscale to the macroscopic level. Whatever the choice, the element of the future must be able to accommodate both physical and chemical aspects of deformation. The answers to such questions will enable the analyst to determine the scope of his models. Such findings will pose challenges to the structural analyst and open up unusually vast and interesting opportunities for research in the mechanics of materials.

We close this section by citing two more examples that illustrate the challenges and opportunities that lie ahead in developing innovative concepts for designing

Figure 9.18. Wind tunnel tests on a dragonfly (Somps and Luttges, 1985). (Courtesy of Springer-Verlag.)

material/structural systems. The first example pertains to dragonflies that continue to challenge our imagination in the context of unsteady aerodynamics of flight. Dragonflies exhibit supermaneuverability, hovering at one instant and darting sideways and backwards at the next instant. They are known to fly at speeds of up to 14 m/s and to lift up to 15 times their own weight. These capabilities are derived not by relying on the smooth flow of air over their wings, but by a process of precise control and exploitation of unsteady aerodynamic forces. In fact, "calculations based on steady-state aerodynamics do not produce the lift values necessary to counterbalance the weight of the insect" (Somps and Luttges, 1985). The front pair of dragonfly wings executes vibration and generates vortices, as shown in Fig. 9.18, and the rear pair of wings, vibrating at a different phase, captures the energy in these vortices resulting in an enhanced flow speed on the rear wings. This quick, continuous and precise control of vibratory motion provides additional lift and allows the insect to maneuver at will. Furthermore, the dragonfly "appears able to readily switch between the use of unsteady flows and the use of more conventional steady-state aerodynamics," using the latter for gliding (Luttges, 1989). The detailed processes that influence this incredible capability are not fully understood yet and therefore the challenge continues. Mimicking in this context requires unconventional approaches. The implications of putting to use, in a controlled way, the unsteady aerodynamics of vibrating components, rather than avoiding them, can be enormous.

The second example comes from a book entitled *Two Years Among New Guinea Cannibals* (Pratt, 1906). It appears the natives of New Guinea appreciated biological materials and made use of them. They would set up in the forest a bamboo pole, one end of which was bent in the shape of a large circle, and leave it there overnight. By the next morning, a local spider had covered the circular space completely with its web, and

Figure 9.19. Fishing net (Pratt, 1906).

the natives used this curious net to fish (Fig. 9.19). Equally remarkable and somewhat more scientific information on spiders is found in a letter by A. G. Courtellemont that appeared in the September 1, 1900 issue of *Scientific American*. It appears the natives of the great island of Madagascar essentially harvested spiders by drawing the silk from them and making splendid fabrics. Spiders are held in the so-called guillotine reeling boxes in groups of one or two dozen, as shown in Fig. 9.20. The spiders were mounted in such a way that their abdomens emerged, and the Malagash girls who performed this delicate operation touched the end of the abdomens of the prisoners and gently withdrew with the finger and carried along in a single bundle the twelve or twenty-four threads to a hook, as shown in Fig. 9.21. The reeling then started, producing a "thread of gold which could not be more brilliant or of a purer yellow."

That was nearly a hundred years ago. Currently research on spider silk is being carried out at Cambridge, Massachusetts, by Dr. Nick Ashley, who says that "spider silk is

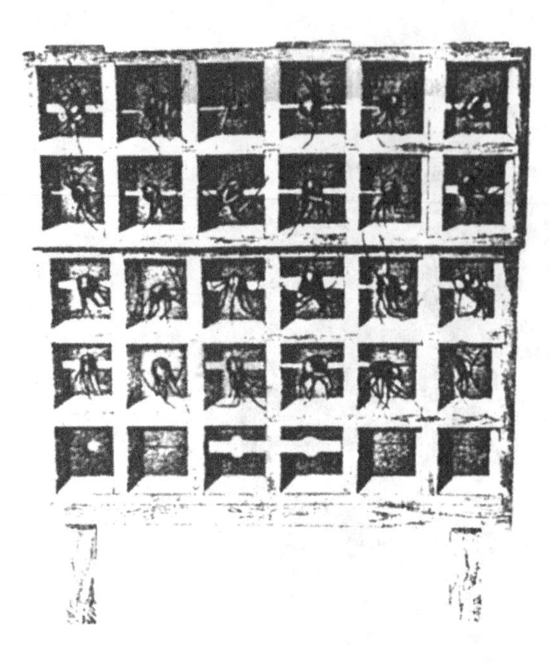

Figure 9.20. Spiders in guillotine reeling-boxes (Courtellemont, 1900). (Courtesy of *Scientific American.*)

comparable in tensile strength to aramid fibers or high strength nylon" (Bacon, 1989). Technological advances in this century have allowed researchers to isolate the gene that produces the silk, and the sequence of instructions is spliced into a host bacterium from which the silk is drawn. In this country, research on spiders is currently active at the Army R&D laboratory at Natick. Claims are made that superior fibers can be developed from this source that "could be used in industrial composites for the automotive and aerospace industries" and that "this silk has rubber like springiness and ability to suspend weights that would snap equivalent strands of steel" (Bacon, 1989).

Figure 9.21. The reeling apparatus (Courtellemont, 1900). (Courtesy of *Scientific American.*)

9.6 Summary

Biomimetics aims at revolutionizing the process of structural design by developing a thorough understanding of the processes that natural systems employ to build strong and durable structures. The scope of such a study is both enormous and fascinating, whether it involves the wing of a butterfly or the leaf of a lily. It requires the combined talents and expertise of the broadest array of scientific and engineering disciplines.

In regard to why the engineering community has not capitalized more on the experience of a half billion years or so of evolutionary optimization that is available from a study of living organisms, we find Professor Gordon's explanation most interesting and revealing. "The mechanical problems of plants and animals are not trivial; they are of quite extraordinary difficulty and complexity .... But, almost invariably, living things are so successful in solving these structural problems that we do not notice how they do it. For nothing attracts less curiosity than total success" (Gordon, 1980). Our efforts to broadcast this "total success" are aimed at attracting the attention of all disciplines contributing to the production of smart structures.

Mimicking a biological system in its totality is neither feasible nor, fortunately, necessary. However, structural design based on the principles optimized in living organisms can lead to innovative solutions to long-standing problems. As noted in the discussion on wood, the increased toughness was obtained with only 10% of the cells buckling. One wonders about the possibility of increasing the percentage of cells undergoing buckling in order to achieve enormous increases in toughness in man-made composites. In a similar manner, natural systems provide us with guidance to design and develop structures with superior strength/weight ratios, increased adaptability, enhanced reliability, and built-in repairability so that future structural systems can truly be considered "smart." We must, however, first unravel and understand these basic principles. Then one would hope that such an understanding when combined with the experience gained in the development of both organic and inorganic composites should lead to material/structural concepts far superior to the most advanced currently available.

For centuries, man has been preoccupied with the notion of "controlling" nature. Biomimetics advocates that more emphasis should be placed on learning from nature by studying the function and architectural order in widely diverse systems such as bone, muscle, antler, cuticle, mollusks, birds, bats, elephants, trees, and leaves. Natural systems hold the key to an inexhaustible supply of insights and concepts that can form the basis for, and serve as a rich source of, innovation that has the potential to bring about significant breakthroughs needed for progress in the analysis, design, and development of smart structures.

BIBLIOGRAPHY

Attenborough, D. 1990. *Trials of Life*. Boston: Little, Brown & Co.

Bacon, M. 1989. A tangled web. *Materials Edge* 32.

Barth, F. G. 1978. Slit sense organs: Strain gauges in the arachnid exoskeleton. *Symp. Zool. Soc.*, London, 42:439–448.

Bennet-Clark, H. C. 1975. The energetic of the jump of the locust. *Journal of Experimental Biology* 63:53–83.
Chaplin, R. C., J. E. Gordon, and G. Jeronimidis. 1983. U.S. Patent No. 4,409,274, 1983.
Cowin, S. C. 1990. Structural adaptation of bones. *Applied Mechanics Reviews* 43(5, Part 2).
Currey, J. D. 1977. Mechanical properties of mother of pearl in tension. *Proceedings of the Royal Society* (B), 196:443–463.
Currey, J. D. 1980. Mechanical properties of mollusk shell. In *The Mechanical Properties of Biological Materials Symposium of the Society of Experimental Biology* 34 (ed. JFV Vincent and J. D. Currey): 75–97, Cambridge: Cambridge University Press.
Currey, J. D. 1983. Biological composites. In *Handbook of Composites*, 4. New York Elsevier Science Publishers.
Currey, J. D. 1990. Private communication.
Currey, J. D. 1991. Private communication.
Currey, J. D., and J. D. Taylor. 1974. The mechanical behavior of some molluscan hard tissues. *Journal of Zoology* 173:395–406.
Evans, A. G. 1990. Perspective on the development of high-toughness ceramics. *Journal of American Ceramic Society* 73(2):187–206.
Evans, A. G. 1991. Private communication.
Gordon, J. E. 1980. Biomechanics: The last stronghold of vitalism. In *The Mechanical Properties of Biological Materials Symposium of the Society of Experimental Biology* 34 (ed. JFV Vincent and J. D. Currey): 1–11, Cambridge: Cambridge University Press.
Gordon, J. E. 1987. *The Science of Structures and Materials*. Scientific American Books, San Francisco.
Gordon, J. E., and G. Jeronimidis. 1980. Composites of high work of fracture. *Philosophical Transactions of the Royal Society of London* A 294:545–550.
Hadley, N. F. 1986. The arthropod cuticle. *Scientific American* 255(1):104–112.
Jackson, A. P., J. F. V. Vincent, and R. M. Turner. 1988. The mechanical design of nacre. *Proceedings of the Royal Society of London* B 234:415–440.
Jackson, A. P., J. F. V. Vincent, and R. M. Turner. 1990. Comparison of nacre with other ceramic composites. *Journal of Materials Science* 25:3173–3178.
Jeronimidis, G. 1980a. The fracture behavior of wood and the relations between toughness and morphology. *Proceedings of the Royal Society of London* B 208:447–460.
Jeronimidis, G. 1980b. Wood, one of nature's challenging composites. In *The Mechanical Properties of Biological Materials Symposium of the Society of Experimental Biology* 34 (ed. JFV Vincent and J. D. Currey):169–182, Cambridge: Cambridge University Press.
Jeronimidis, G. 1991. Private communication.
Kastelic, J., and E. Baer. 1980. Deformation in tendon collagen. In *The Mechanical Properties of Biological Materials Symposium of the Society of Experimental Biology* 34(ed. JFV Vincent and J. D. Currey):397–435, Cambridge: Cambridge University Press.
Katsamanis, F., L. Wei, D. Raftopoulos, and F. Saul. 1990. The Hopkinson bar technique for the determination of mechanical properties of human cortical bone. In *Advances in Bioengineering* (ed. S. Goldstein), Vol. ASME BED 17.
Kitchener, A. C. 1987. Fracture toughness of horns and a reinterpretation of the horning behavior of bovids. *Journal of Zoology* 213:621–639.
Kitchener, A. C. 1988. An analysis of the forces of fighting of the blackbuck and the bighorn sheep and the mechanical design of horns of bovids. *Journal of Zoology* 214:1–20.
Laraia, V. J., and A. H. Heuer. 1989. Novel composite microstructure and mechanical behavior of mollusk shell. *Communications of the American Ceramic Society* 72(11):2177–2179.

Laraia, V. J., and A. H. Heuer. 1991. Private communication.

Luttges, M. W. 1989. Accomplished insect fliers. *Frontiers in Experimental Fluid Mechanics* (ed. M. Gad-el-Hak) 46:429–456, Springer-Verlag, Berlin.

Ma, M., K. Vijayan, A. Hiltner, E. Baer, and J. Im. 1990. Thickness effects in microlayer composites of polycarbonate and polystyrene-acrylonitrile. *Journal of Materials Science* 25:2039–2046.

Pietsch, T. W., and D. B. Grobecker. 1990. Frogfishes. *Scientific American* 262(6):96–103.

Pratt, A. E. 1906. *Two Years among New Guinea Cannibals*. London: Sealy & Co. Ltd.

Prewo, K. M., and J. J. Brennan. 1982. Silicon carbide fiber-reinforced glass-ceramic-matrix composites exhibiting high strength and toughness. *Journal of Materials Science* 17:1201–1206.

Sadananda, R. 1991. A probabilistic approach to bone fracture analysis. *Journal of Materials Research* 6(1):202–206.

Sarikaya, M., K. E. Gunnison, M. Yastrebi, and I. A. Aksay. 1990. Mechanical property-micro-structural relationships in abalone shell. *Materials Research Society Symposium Proceedings* 174:109–116.

Schiavone, R. C. 1987. Potential applications of biotechnology to aerospace materials. Technical Report TRUDR-TR-87-30, U.S. Air Force Wright Laboratory.

Somps, C., and M. W. Luttges. 1985. Dragonfly flight: Novel uses of unsteady separated flows. *Science* 228:1326–1329.

Srinivasan, A. V., G. K. Haritos, and F. L. Hedberg. 1991. Biomimetics: Advancing man-made materials through guidance from nature. *Applied Mechanics Reviews* 44(11):463–482.

Srinivasan, A. V., G. K. Haritos, F. L. Hedberg, and W. F. Jones. 1996. Biomimetics: Advancing man-made materials through guidance from nature—an update. *Applied Mechanics Reviews* 49(10, Part 2):S194–S200.

Steele, C. R. 1992. Electroelastic behavior of auditory receptor cells. *Biomimetics* 1(1):5–23.

Vincent, J. F. V. 1990. The design of natural materials and structures. *Journal of Intelligent Material Systems and Structures* 1(1):141–146

Vincent, J. F. V. 1991. *Structural Biomaterials*. Princeton, NJ: Princeton University Press.

Vincent, J. F. V., and P. Owers. 1986. The mechanical design of hedgehog spines and porcupine quills. *Journal of Zoology* (A) 210:55–75.

Vogel, S. 1988. *Life's Devices: The Physical World of Animals and Plants*. Princeton, NJ: Princeton University Press.

Wainwright, S. 1980. Adaptive materials: A view from the organism. In *The Mechanical Properties of Biological Materials Symposium of the Society of Experimental Biology* 34 (ed. JFV Vincent and J. D. Currey):437–453, Cambridge: Cambridge University Press.

Weiner, S., and W. Traub. 1991. Crystal organization in rat bone lamellae. In *Workshop on the Design and Processing of Materials by Biomimicking*, Seattle, Washington: Air Force Office of Scientific Research and the University of Washington.

PROBLEMS

1. Given the moduli of elasticity of the major constituents and recognizing that each wood fiber is itself a composite, calculate the degradation in the modulus of elasticity at each level from fiber to cell wall to wood, using the rule of mixtures

and the architecture of the tracheid. Show that the degraded value is 10 GPa, given that the moduli and corresponding volume fractions are as follows:

Cellulose crystallite	250 GPa	0.8
Amorphous region	50 GPa	0.2
Matrix	1 GPa	0.5

2. Considering at least five structural components of interest to you, identify possible analogs from natural structures and make comparisons of properties, mechanisms, design strategies, and principles.

3. In Section 9.2, characteristics of natural structures are presented. They are multifunctionality, hierarchical organization, and adaptability. Assuming that enabling technologies are available to permit design of structures to exploit these characteristics, discuss how you may design a building, a bridge, and a refrigerator.

4. Even though it is not always possible to identify an *exact* biological analog for every structure of engineering interest, analogs at component levels may be useful in designing machinery. For example, although a jet engine may not have a *direct* analog (biologists will argue and point to some self-propelling microorganisms!), many examples in the marine world are found in which fluid is conveyed from a lower pressure region to a higher one, analogous to the function of a fan blade. In this context, select one engineering product you are familiar with and identify one or more analogs at the product level or at component levels.

5. With reference to Fig. 9.6, it was noted that the S2 layer contributes up to 80% of the total thickness of a wood cell and that the helical arrangement of fibers has an important role to play in the fracture mechanism of wood. In this context, explain why it is important to avoid

(a) heavy landing of gliders,

(b) using the back of an oar to stop a rowing boat, and

(c) turning over a diving board after prolonged use.

6. A design strategy adopted by most muscular systems in nature (e.g., tongue, squid tentacles) is large length-to-diameter ratios to achieve large distance advantage with limited muscle contraction. Considering a constant volume cylinder of π units, show that a cylinder whose radius is 1/20th of its length can obtain a force advantage of about 0.13 (long, easy, rapid push) and a distance advantage of 7.9 compared to a cylinder of one unit radius and one unit length when an outer circumferential muscle contraction of one unit in girth occurs. Compare the force and distance advantages.

APPENDIX A

Selected Topics from Structural Dynamics

A.1 Introduction

When a structure or structural component is subjected to time-dependent forces, the response of the structure or component can be time-dependent. Such a response is characterized as vibratory response. It is important to understand the vibratory characteristics of structures or structural components because their structural integrity (and therefore their service life) depends on their ability to withstand the time-dependent stress states. Vibration can occur in any body possessing mass and elasticity. Therefore, most engineering machines and structures can be susceptible to vibratory motion. Some of the common examples of structures that are generally known to respond to time-dependent forces are airplanes, helicopters, ships, submarines, buildings, bridges, railways, automobiles, engines, a variety of rotating machinery such as pumps and compressors, musical instruments, and a variety of machines and devices that are commonly used in transportation, manufacturing, and other industries. These structures and machines possess mass and elasticity and, therefore, they are likely to be responsive to time-dependent force systems. Practically everything we know of can vibrate – whether or not such vibration is noticeable or even important from the point of view of its structural integrity.

The structures we discussed above are products of engineering design. As such, they are "sized" to withstand forces to which they are subjected to in the course of their service life. In that context, it becomes important to design them such that the total maximum stress (sum of steady and dynamic) experienced anywhere in the structure is within the limits of capability of the material used in its design. In the world of serious competition among industries, products that succeed in the marketplace will be those that have used the material in an optimum way so that component volumes and weights are held to a desired minimum even as the performance and service life are required to be high.

Some of the commonly used definitions and terminology follow.

A.2 Linear and Nonlinear Vibration

The principle of superposition holds in linear behavior. Thus, the response of a structure to a system of applied forces can be obtained by summing the responses due to each. As the response is linearly related to excitation, doubling the magnitude of the force of excitation results in doubling of the response. Nonlinearity implies that no such simple relationship exists between the stimulus and the response.

A.3 Free and Forced Vibration

When vibration takes place without any external stimulus acting continuously, the resulting response is termed free vibration response. In most cases, some stimulus is necessary to begin the process but vibration may continue even after the removal of the excitation. The extent of time in which this continues depends upon the level of damping in the system. Free vibration takes place under the action of forces inherent in the system itself. In a mathematical sense, free vibration analysis pertains to the solution of homogeneous differential equations. Forced vibration occurs when a system of time-dependent forces is applied to a structure or structural component.

A.4 Natural Frequencies

Frequency of vibration is a measure of the extent of repetition of vibratory behavior. How frequently does the pattern of vibration repeat itself over a given interval of time? For example, is it less frequent, more frequent, or extremely frequent in an interval of time such as for example every second? Frequency of vibration is expressed as so many cycles (repetition) per second. As mentioned earlier, any component possessing mass and elasticity can vibrate and such vibration can take place at a number of discrete frequencies and continue without a continuous application of external forces. These discrete frequencies characterize a structure dynamically and are known as natural frequencies. The natural frequencies of a vibrating system represent one of the most fundamental dynamic characteristic and are the essential basic information needed to determine the suitability of a design. In a mathematical sense, the natural frequencies are the eigenvalues of a vibrating system.

A.5 Resonance and Damping

At each of its natural frequencies, a structure is capable of experiencing large response (strain, displacement, acceleration, etc.) and is said to resonate. The extent of these resonant responses is limited entirely by the nature and extent of inherent forces of dissipation. The latter are known as forces of damping and are essential in every structural component to prevent responses from escalating to dangerous levels. It is the presence of damping forces that prevents vibratory failure, and therefore damping represents another important dynamic characteristic.

A.6 Degrees of Freedom

Degrees of freedom represent the extent of flexibility that needs to be built into a mathematical model to represent a vibrating component in order to conduct a dynamic analysis. Theoretically every structure/component possesses infinite degrees of freedom and therefore infinite natural frequencies. Usually not all of these frequencies are important because the range of interest is confined to the range of frequency content of the force system that the structure has to endure. Clearly, the mathematical representation need include only that range of frequencies that is physically significant, and therefore it is usual to limit the number of degrees of freedom in modeling. Thus, both the physical and mathematical requirements lead to discussions of finite-degree-of-freedom systems.

A.7 Modes of Vibration

When a structure represented by certain degrees-of-freedom is analyzed for vibration, the vibratory responses at a natural frequency, at different locations on the structure at a given instant of time, display a pattern of deformation termed as a mode of vibration. There is always a mode of vibration corresponding to a natural frequency. Modes of vibration along with corresponding natural frequencies and damping constitute another important dynamic characteristic. In a mathematical sense, the mode shapes correspond to eigenvectors of a vibrating system corresponding to the eigenvalues.

A.8 Periodic Vibration

Vibration repeated at equal intervals of time is known as periodic vibration, and the interval of time is known as the period of vibration. If $X(t)$ represents a vibratory motion that repeats itself every interval of time τ, then $X(t+\tau) = X(t)$.

A.9 Harmonic Vibration

Harmonic motion is a special case of periodic vibration in which motion takes place at a single frequency. It can be represented by a sine or cosine function of time. For example, if the period of harmonic vibration is τ, then the time interval t within the period can be represented as a fraction of t/τ. Thus, $t/\tau = 0$ may represent the beginning of motion and $t/\tau = 1$ may represent the completion of one cycle of vibration. Thus, the motion can be conveniently expressed as $X = A \sin 2\pi(t/\tau)$. The quantity $1/\tau$ represents the frequency of vibration in cycles/second (f), while $2\pi/\tau$ is known as circular frequency, commonly denoted by ω. The expression $\omega = 2\pi f$ can be used to represent harmonic vibration as a projection on a straight line of a point moving on a circle at a constant speed ω.

A.10 Wave Form/Fourier Analysis

A wave form representing vibratory motion represents time history of motion measured (or calculated) at a point on a structure. If the wave form is of a general nature but is periodic, then it can be considered as a superposition of a number of contributing harmonics each of which is a vector, and a combination of all of which constitutes the wave form.

A.11 Statistical Measures/Mean Value

$$\bar{X} = \lim_{T\to\infty} \frac{1}{T} \int_0^T X(t)\, dt. \tag{A.1}$$

The integral represents the total area of the curve within the period of vibration. Thus the mean value of a pure sine wave is zero. Mean value for a half-sine is $2/\pi$.

A.12 Mean Square

The square of vibratory displacement is an important parameter in vibration because energy stored in a structure is proportional to square of the displacement. Thus, a measure of energy in a structure can be obtained by calculating

$$\bar{X}^2 = \lim_{T\to\infty} \frac{1}{T} \int_0^T X^2(t) = \frac{A^2}{2} \tag{A.2}$$

for a sinusoidal motion of amplitude A.

A.13 Root Mean Square

The square root of the mean square value is also used as a measure of vibration levels and is defined as $\sqrt{\bar{X}^2}$. Therefore, for a sinusoidal vibration, the RMS is $0.707A$.

APPENDIX B

Selected Topics from Automatic Control

B.1 Introduction

Several topics from automatic control theory are reviewed here and some general conclusions are drawn, but it must be realized that only the most superficial treatment of the subject can be given in this space. Some advanced topics – stochastic control and adaptive control, for example – have been omitted, but introductory treatments and pointers to further information may be found in the references given here and at the end of Chapter 8.

The student or practitioner approaching the subject of smart structures will in general have some knowledge of either control theory or structural mechanics, but it is rare that he has a working knowledge of both fields. The more common need seems to be for a structural engineer to learn something of automatic control, and it is this need we will attempt to address in this appendix. The controls specialist is, however, cautioned not to regard a structure as "just another plant."

B.2 Classification of Systems

In discussing dynamic systems, it is often useful to identify characteristics that reflect their complexity. Of particular interest are features that indicate how a system must be modeled or which control methods might be appropriate for it. As one might expect, the analysis and design techniques applicable to the simplest categories of systems are the most mature; fortunately, these are adequate for many realistic smart structures.

One of the most fundamental distinctions is between linear and nonlinear systems. Linearity implies that a system's output is proportional to its input, and thus that the output is zero in the absence of any input. Furthermore, the response to the sum of two (or more) simultaneous inputs is given by the sum of the responses to those inputs occurring singly. If an arbitrary system is described by the function H, and u_1 and u_2 are two inputs to this system, the system is linear if and only if

$$H(\alpha_1 u_1 + \alpha_2 u_2) = \alpha_1 H(u_1) + \alpha_2 H(u_2), \tag{B.1}$$

where α_1 and α_2 are constants. Linearity is a very strong condition, and one that allows a large body of highly developed mathematics to be brought to bear more or less automatically, but it is certainly not essential to show or assume a system to be linear in order to proceed with its analysis. The literature on the control of nonlinear systems is voluminous, and applications specifically to components of smart structures are beginning to appear.

Once it is determined whether a system is linear, it is usually desired to know if its parameters or mathematical structure are changing with time – that is, is the system *time varying* or *time invariant*. Changes in the plant (the system or structure to be controlled) over time can be due to simple aging of components, or to environmental factors, such as temperature fluctuations, or they can be inherent in the system's operation, such as when a structure's geometry changes as it is deployed. Often these variations will be slow compared to the dynamic phenomena the control system is designed to deal with, but they may make it impossible to use a single model of the plant for all phases of its operation.

Although the notions of linearity and time-invariance are independent, it is common to see a system described as *linear time varying* (LTV) or *linear time invariant* (LTI). The latter class of systems is generally regarded as the simplest to analyze and the easiest for which to design a controller. In fact, some introductory texts are devoted to a subset of LTI systems, those with a single input and a single output, commonly represented by the scalar functions $u(t)$ and $y(t)$, respectively. Somewhat more complex, and often modeled and analyzed using different mathematical tools, are LTI systems with multiple inputs or outputs. Which category a system falls into is denoted by abbreviations like SISO (single-input, single-output) and MIMO (multi-input, multi-output). Thus, SISO LTI systems may be considered the simplest and most tractable; indeed, one approach to modeling a given system is to ask what complexities need be added to this starting point. Unless stated otherwise, we shall confine our treatment here to LTI systems.

B.3 Classical Control

In *classical control*, an approach where the physical nature of the model, the input, and the output remain evident during the analysis, the plant is represented by a linear differential equation describing variables such as position and rate. This equation is usually written first in the time domain, and is of low order (especially compared to some structural dynamic models). The controller monitors the output of the plant and adjusts the net plant input to bring about a desired condition or action, and is likewise designed in terms of physical quantities or parameters.

Let the dynamics of the plant be represented by the function $g_p(t)$ and the input and output by $u(t)$ and $y(t)$, respectively. (The notation used in this appendix follows that of Kuo, 1995.) The function $g_p(t)$ is the *impulse response* of the plant, that is, the solution to its differential equation when forced by the impulse $\delta(t)$ and subject to zero initial conditions. This simple system can be depicted schematically as shown in Fig. B.1. The mathematics implied by this block structure is most easily understood

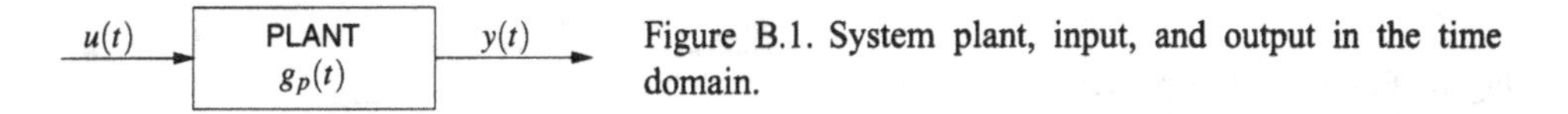

Figure B.1. System plant, input, and output in the time domain.

in the frequency domain, reached through Laplace transformation of the governing differential equation and other functions of time. (*Frequency* here must be understood as the complex variable $s = \sigma + j\omega$, the imaginary part of which may be regarded as a physical frequency in radians per second.) Following this transformation, the input $U(s)$ can simply be multiplied by the plant *transfer function* $G(s)$, which is the Laplace transform of the impulse response $g(t)$, to obtain the output $Y(s)$. The block diagram, Fig. B.2, changes little, but it now indicates more directly how to obtain $Y(s)$.

Similarly, a feedback block may be described by a differential equation with transfer function $H(s)$. Its input is the plant output $Y(s)$ and its output, denoted $B(s)$, is part of the plant input $U(s)$. The remainder of $U(s)$ is an external command or reference signal, $R(s)$. The relationships among these signals and subsystems are shown in Fig. B.3, where the minus sign at the junction of $R(s)$ and $B(s)$ reflects the choice of negative feedback,

$$U(s) = R(s) - B(s). \tag{B.2}$$

This choice of sign is conventional, and follows from the observation that negative feedback is generally necessary for the closed-loop system to be stable.

An advantage of working in the frequency domain is that signals and the structure of block diagrams can be manipulated algebraically. The outputs of parallel blocks are added, while the transfer functions of blocks in series are multiplied. For the canonical feedback system of Fig. B.3, the output may be calculated as

$$Y(s) = G(s) - U(s) \tag{B.3a}$$

$$= \frac{G(s)}{1 + G(s)H(s)} R(s). \tag{B.3b}$$

The *closed-loop transfer function* $M(s)$, the ratio of the output of this system to the input command, is thus

$$M(s) = \frac{Y(s)}{R(s)} = \frac{G(s)}{1 + G(s)H(s)}. \tag{B.4}$$

The corresponding impulse response of the closed-loop system could be found by inverse Laplace transformation of eq. (B.4), but this is rarely used in practice.

Three fundamental simplifications used frequently in block diagram algebra are shown in Fig. B.4. By applying these alternately to blocks in series and in parallel and to feedback loops, one can reduce systems of arbitrary complexity to as few blocks as

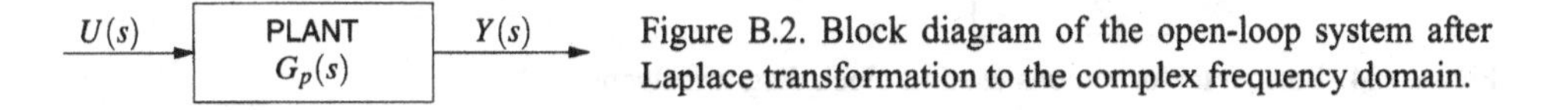

Figure B.2. Block diagram of the open-loop system after Laplace transformation to the complex frequency domain.

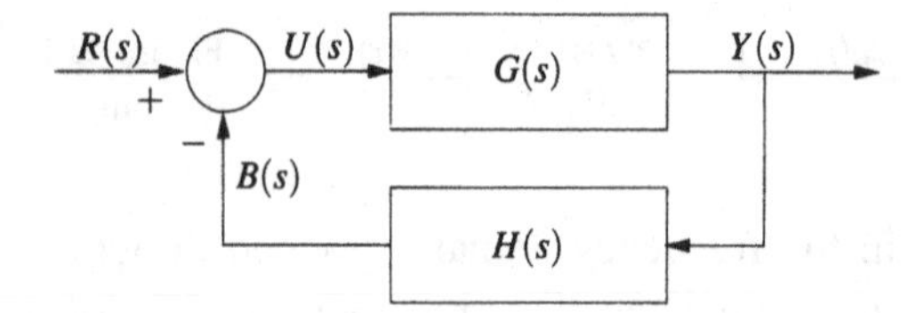

Figure B.3. Closed-loop system. The minus sign indicates negative feedback of the signal $B(s)$.

desired. The calculations involved can be tedious, however, and there will likely be some loss in ease of physical interpretation of the mathematics.

Because the system transfer function $M(s)$ embodies the effects of the controller as well as the dynamics of the plant, it forms the basis of much of classical control theory. Most analyses are concerned with the stability of the closed-loop system, the rate at which it responds to a command input or disturbance, or the transient or steady-state errors in its response. Corresponding design techniques allow one to examine how these system characteristics vary with the structure and parameters of the controller.

Most results in classical control are calculated and interpreted in the frequency domain, and may be represented in the complex plane. However, some important information about the time response of the system can often be obtained by using the initial and final value theorems, which state, respectively,

$$\lim_{t \to 0} f(t) = \lim_{s \to \infty} sF(s) \tag{B.5}$$

and

$$\lim_{t \to \infty} f(t) = \lim_{s \to 0} sF(s), \tag{B.6}$$

where, as suggested by the notation, $F(s)$ is the Laplace transform of $f(t)$. Application of the final value theorem requires that the system be stable. Use of these formulae allows the very short- and long-time responses of a system to be calculated directly by defining $F(s)$ as the product of the system transfer function and the s-domain representation of the input signal of interest, i.e., $F(s) = G(s)U(s)$.

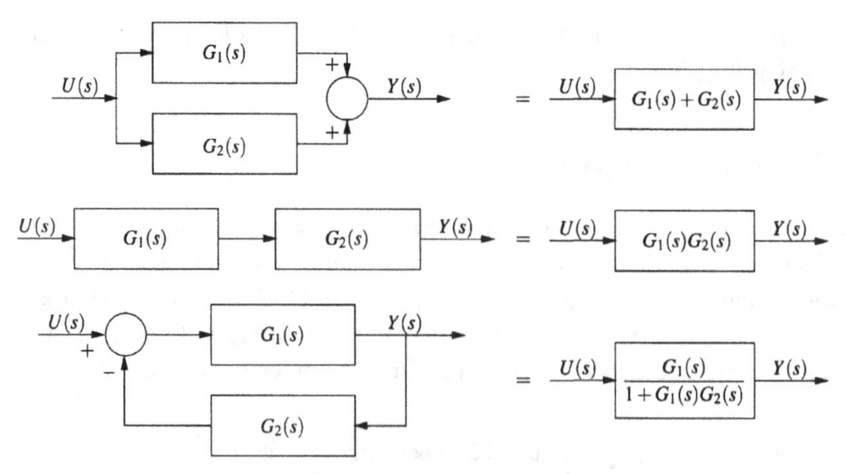

Figure B.4. Basic transformations used in block diagram algebra.

The stability of a system concerns its response to disturbances, and is often examined in terms of the system's behavior following a small perturbation from an equilibrium condition or operating point. If its response returns to equilibrium after the perturbation, the system is stable; if, instead, its response grows with no further input, the system is unstable. Clearly, most engineering systems, and structures in particular, must be stable if they are to be at all reliable or useful. Although most passive structures are inherently stable (an important exception being structures subject to buckling), the stability of active systems, including closed-loop smart structures, cannot be taken for granted. For purposes of illustration, consider the simple first-order system described by the equation

$$\dot{x}(t) + ax(t) = u(t), \tag{B.7}$$

where a is real. The homogeneous solution of this differential equation is simply $x_h(t) = Ce^{-at}$ where C is a constant, from which it may be seen that following a disturbance the response $x(t)$ will return exponentially to zero (if $a > 0$) or diverge exponentially (if $a < 0$).

Fortunately, it is usually easy to test the stability of a system represented by a transfer function. Associated with any linear system is a characteristic equation, the roots of which reveal quantitative information about the stability of the system. This characteristic equation is obtained simply by setting the denominator of the system's transfer function equal to zero; this denominator is often a polynomial of low to moderate degree. The roots of the characteristic equation may be real or complex, the latter occurring in conjugate pairs, and will generally depend on parameters of the controller that may be selected by the designer. Thus, the characteristic roots can be manipulated to some extent during the design process.

If all the roots of the characteristic equation are in the left half of the complex plane (i.e., $\operatorname{Re} s_i < 0$ for all roots s_i), the system is stable, and it will return asymptotically to equilibrium following a perturbation. If any of the roots are on the imaginary axis, $\operatorname{Re} s_i = 0$, the system is marginally stable, and although the disturbance following a perturbation may persist indefinitely, it will not grow in amplitude. Finally, if the characteristic equation has one or more roots in the right half-plane, $\operatorname{Re} s_i > 0$, the system is unstable; the response to a perturbation will grow without bound, exponentially diverging from equilibrium. The transfer function of the system represented by eq. (B.7) is

$$G(s) = \frac{X(s)}{U(s)} = \frac{1}{s + a} \tag{B.8}$$

and hence its characteristic equation is

$$s + a = 0, \tag{B.9}$$

from which we can immediately draw the same conclusion about its stability as was reached above from the time-domain solution of its homogeneous DE. One advantage of working with the characteristic equation, especially for systems of higher order, is that no explicit expression for the response need be found.

Being zeros of its denominator, the roots of the characteristic equation are poles of the transfer function $M(s)$. Zeros of the numerator, on the other hand, are zeros of the transfer function. The locations of these poles and zeros in the complex plane are naturally related to the system's stability and response, and their calculation is a routine part of any classical control analysis or design. The *root locus plot* shows graphically how the poles and zeros move in the complex plane as parameters of the system, typically controller gains (coefficients), are varied.

Another very useful graphical representation of the closed-loop transfer function is the *Bode plot*, where magnitude and phase are plotted against frequency. The input is assumed to be harmonic with frequency ω, and because the system is linear, the output will be of the same frequency but altered in magnitude and phase with respect to the input. Under these conditions, the complex variable s can be replaced by $j\omega$ and we can examine

$$M(j\omega) = M(s)|_{s=j\omega}. \tag{B.10}$$

The magnitude and phase of the steady-state transfer function $M(j\omega)$ (also called the *frequency response function,* as in structural dynamics) are simply

$$|M(j\omega)| = ([\mathrm{Re}\, M(j\omega)]^2 + [\mathrm{Im}\, M(j\omega)]^2)^{1/2} \tag{B.11}$$

and

$$\arg M(j\omega) = \tan^{-1} \frac{\mathrm{Im}\, M(j\omega)}{\mathrm{Re}\, M(j\omega)}. \tag{B.12}$$

Note that while $j\omega$ is purely imaginary, $M(j\omega)$ will in general still be complex valued. The Bode plot thus consists of two curves: a plot of $|M(j\omega)|$ versus ω, and a plot of $\arg M(j\omega)$ versus ω.

The stability of a system can be determined through these graphical methods as well as from the form and coefficients of the characteristic equation. Often, detailed calculations can be avoided by the application of test criteria or by employing approximate graphical construction rules. However, there are many special cases with which one must be familiar, and the reader is urged to consult a thorough reference (for example, Kuo, 1995) before applying these techniques to a particular problem. Such a text will also introduce additional, related methods of analysis.

We come now to the question of how to design the controller itself. Although this can in principle be arbitrarily complex, it is common to consider controllers with output proportional to an error signal, or to the integral or derivative of this error with respect to time. The error is usually defined as the difference between the system response and the desired or commanded response,

$$e(t) = y(t) - r(t) \tag{B.13}$$

or

$$E(s) = Y(s) - R(s). \tag{B.14}$$

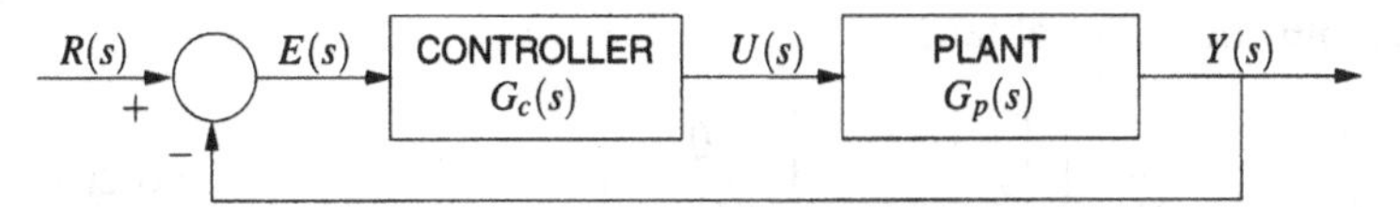

Figure B.5. Structure of a series-compensated closed-loop system. Note the generation of the error signal by unity negative feedback of the plant output.

The controller generates the plant input from this error signal as shown in Fig. B.5. Because both the controller and the plant are in the forward signal path, we denote them by $G_c(s)$ and $G_p(s)$, respectively. This configuration, known as *series compensation,* is one of several possible system block structures. Its popularity is in part due to the ready availability of controller hardware for many applications.

B.4 Modern Control

Although classical control theory lends itself to the manual analysis of simple, lumped-parameter LTI systems, it is sometimes necessary to bring to bear methods of greater mathematical sophistication, or to analyze systems described by many variables. The techniques of *modern* or *state-space* control fill these needs while remaining applicable to simpler problems. In addition to providing tools for the analysis and design of complex control systems, modern control forms the basis for much of the subject of optimal control, discussed below.

The *state* of a system consists of the information needed to completely describe its configuration, such as the position and velocity of each lumped mass of a structure or the voltage across each capacitor, resistor, and inductor in an electrical circuit. One or more ordinary differential equations describing a system may be converted to state-space form by defining as state variables the variables in the original equation plus all their derivatives except the highest. Each higher-order differential equation is replaced by a set of coupled first-order equations in this way. For example, the governing equation of a single-degree-of-freedom mechanical system,

$$m\ddot{x}(t) + c\dot{x}(t) + kx(t) = u(t), \tag{B.15}$$

may be reduced to

$$\dot{x}_1(t) = x_2(t), \tag{B.16}$$

$$m\dot{x}_2(t) + cx_2(t) + kx_1(t) = u(t) \tag{B.17}$$

by defining

$$x_1 = x, \tag{B.18}$$

$$x_2 = \dot{x}_1. \tag{B.19}$$

In general, an nth-order DE can be reduced to n first-order equations.

Modern control theory draws heavily on results from linear algebra, a connection that is greatly facilitated by rewriting the set of first-order scalar DEs as a matrix equation. For the example above, we obtain

$$\begin{bmatrix} \dot{x}_1(t) \\ \dot{x}_2(t) \end{bmatrix} = \begin{bmatrix} 0 & 1 \\ -k/m & -c/m \end{bmatrix} \begin{bmatrix} x_1(t) \\ x_2(t) \end{bmatrix} + \begin{bmatrix} 0 \\ u(t)/m \end{bmatrix} \tag{B.20}$$

or simply

$$\dot{\mathbf{x}}(t) = \mathbf{A}\mathbf{x}(t) + \mathbf{u}(t). \tag{B.21}$$

It is important to realize that, while this example is extremely simple, state-space control methods are applicable to systems of larger dimension; the necessary theoretical and computational tools can usually be developed without explicit consideration of the size of the problem to be addressed (although the extremely large models sometimes encountered in structural dynamics may require special treatment, as explained below).

Let us now consider the more general case of a system with state variables $x_1, \ldots, x_n$ stacked into the vector $\mathbf{x}$. Because there may be more than one input, $\mathbf{u} = (u_1 \ \cdots \ u_p)^T$ likewise becomes a vector of length p equal to the number of inputs. The state equation for the system is then

$$\dot{\mathbf{x}}(t)_{(n\times 1)} = \mathbf{A}_{(n\times n)}\mathbf{x}(t)_{(n\times 1)} + \mathbf{B}_{(n\times p)}\mathbf{u}(t)_{(p\times 1)}. \tag{B.22}$$

The elements of the matrix $\mathbf{A}$ are determined by the system dynamics and those of $\mathbf{B}$ by the manner in which the state equations are driven by the inputs. To accompany this, we have a second matrix equation,

$$\mathbf{y}(t)_{(q\times 1)} = \mathbf{C}_{(q\times n)}\mathbf{x}(t)_{(n\times 1)} + \mathbf{D}_{(q\times p)}\mathbf{u}(t)_{(p\times 1)}, \tag{B.23}$$

which expresses the output(s) $\mathbf{y}$ of the system in terms of its state and inputs. In general, the number of states will exceed the number of outputs (i.e., $n > q$).

When the system under consideration is LTI, it is possible to take the Laplace transforms of the state and output equations. Ignoring initial conditions as before, this results in

$$\mathbf{X}(s) = (s\mathbf{I} - \mathbf{A})^{-1}\mathbf{B}\mathbf{U}(s), \tag{B.24}$$

$$\mathbf{Y}(s) = \mathbf{C}\mathbf{X}(s) + \mathbf{D}\mathbf{U}(s). \tag{B.25}$$

Substituting for $\mathbf{X}(s)$ in the second of these equations leads to

$$\mathbf{Y}(s) = \mathbf{G}(s)\mathbf{U}(s) \tag{B.26}$$

where

$$\mathbf{G}(s)_{(q\times p)} = \mathbf{C}(s\mathbf{I} - \mathbf{A})^{-1}\mathbf{B} + \mathbf{D}. \tag{B.27}$$

This is called the system's *transfer function matrix*, and it is analogous to the scalar transfer function $G(s)$ relating the input and output of a SISO system.

Earlier we saw that when a system is described by a scalar transfer function we can obtain its characteristic equation by setting the denominator of that transfer function equal to zero. The same characteristic equation can be found from the state variable representation of the system, without necessarily calculating its transfer function matrix. In eq. (B.27), the denominator of the right-hand side will result from the inversion of the factor $(s\mathbf{I} - \mathbf{A})$. Setting that denominator to zero, we immediately obtain the characteristic equation

$$\det(s\mathbf{I} - \mathbf{A}) = 0. \tag{B.28}$$

It may now be seen that the values of s satisfying eq. (B.28) are also the eigenvalues $\lambda_1, \ldots, \lambda_n$ of the matrix $\mathbf{A}$. The locations of these eigenvalues in the complex plane indicate the stability of the system as previously described (e.g., if $\operatorname{Re} \lambda_i < 0$ for $i = 1, \ldots, n$, the system is stable). Corresponding to each eigenvalue λ_i there is an eigenvector $\mathbf{p}_i$ of $\mathbf{A}$ satisfying the equation

$$(\lambda_i \mathbf{I} - \mathbf{A})\mathbf{p}_i = \mathbf{0}. \tag{B.29}$$

Some complexity may be introduced in the event $\mathbf{A}$ is not symmetric and has one or more repeated eigenvalues, in which case generalized eigenvectors $\mathbf{p}_i$ can be computed from a set of auxiliary matrix equations. Details of this calculation are omitted here in the interest of simplicity.

Once we have a representation of a system in terms of state and output equations, it is frequently advantageous to find an equivalent representation in terms of a different set of state variables. This is similar to how conversion to the proper choice of generalized coordinates can simplify analysis in classical mechanics. Such a change of coordinates is conveniently defined via matrix multiplication,

$$\mathbf{x} = \mathbf{P}\bar{\mathbf{x}}, \tag{B.30}$$

where $\mathbf{P}$ is a prescribed n-square matrix. Substituting this into the original state equation gives

$$\mathbf{P}\dot{\bar{\mathbf{x}}} = \mathbf{A}\mathbf{P}\bar{\mathbf{x}} + \mathbf{B}\mathbf{u}, \tag{B.31}$$

and after premultiplying by $\mathbf{P}^{-1}$,

$$\dot{\bar{\mathbf{x}}} = \bar{\mathbf{A}}\bar{\mathbf{x}} + \bar{\mathbf{B}}\mathbf{u}, \tag{B.32}$$

where

$$\bar{\mathbf{A}} = \mathbf{P}^{-1}\mathbf{A}\mathbf{P}, \tag{B.33}$$

$$\bar{\mathbf{B}} = \mathbf{P}^{-1}\mathbf{B}. \tag{B.34}$$

Similarly, for the output equation,

$$\mathbf{y} = \bar{\mathbf{C}}\mathbf{P}\bar{\mathbf{x}} + \bar{\mathbf{D}}\mathbf{u} \tag{B.35}$$

and hence

$$\bar{\mathbf{y}} = \bar{\mathbf{C}}\bar{\mathbf{x}} + \bar{\mathbf{D}}\mathbf{u}, \tag{B.36}$$

where

$$\bar{\mathbf{y}} = \mathbf{y}, \tag{B.37}$$
$$\bar{\mathbf{C}} = \mathbf{CP}, \tag{B.38}$$
$$\bar{\mathbf{D}} = \mathbf{D}. \tag{B.39}$$

The dynamics with respect to the new coordinates $\bar{\mathbf{x}}$ are embodied in the system matrix $\bar{\mathbf{A}}$ obtained through the similarity transformation (B.33). Because the underlying physical system has not changed, one would expect the matrices $\mathbf{A}$ and $\bar{\mathbf{A}}$ to have some properties in common. In fact, it can be shown that a similarity transformation preserves the characteristic equation (and thus the eigenvalues), the eigenvectors, and the transfer function matrix of $\mathbf{A}$.

The choice of the matrix $\mathbf{P}$ depends upon the desired form (matrix structure) of $\bar{\mathbf{A}}$. If $\bar{\mathbf{A}}$ can be rendered sparse, for example having only one nonzero diagonal and one nonzero row or column, characteristics of the system may be more readily deduced. Among the most useful transformations are those that produce the *controllable* and *observable canonical forms* of the state equation, where the transformation matrix is found from $\mathbf{A}$ and, respectively, $\mathbf{B}$ or $\mathbf{C}$, and the *Jordan form*, where the transformation matrix consists of the (generalized) eigenvectors of $\mathbf{A}$ and the resulting $\bar{\mathbf{A}}$ is diagonal (or nearly so).

The notions of controllability and observability have intuitive meanings as well as rigorous mathematical definitions. Although the controllability of a given system may be tested by using criteria closely related to the transformations to the corresponding canonical forms, for the purposes of this text it will suffice to convey the physical meaning of these concepts. Simply put, a system is *completely controllable* if every state variable in $\mathbf{x}$ is affected by the control $\mathbf{u}$, and *completely observable* if every state variable affects the output $\mathbf{y}$. Note that some state variables may be controllable or observable, and others not; such a system is said to be uncontrollable or unobservable, respectively. These definitions, based on state variables, can be recast in terms of the input-output relationship described by a transfer function (see, e.g., Kuo, 1995).

Whereas previously we used block diagram algebra to incorporate an output-feedback controller into a model of a SISO system, in modern control the quantity fed back is often some combination of the state variables. The state vector $\mathbf{x}$ is augmented by variables describing the state of the controller, and the various coefficient matrices are expanded as necessary to reflect the dynamics of the new subsystems. The analytical techniques described above are still applicable; for example, the stability of a closed-loop system can be determined by examining the eigenvalues of the augmented $\mathbf{A}$ matrix.

B.5 Optimal Control

The concern of classical and modern control theory is chiefly to track command inputs acceptably fast while rejecting disturbances. This end is accomplished through some form of feedback, typically of either the system output or the system state. Performance is specified through criteria that must be met by a successful design. *Optimal control* differs from this in that an objective is defined and a system then designed to achieve this objective in the best way possible, subject to given constraints. "Best" here may mean, for example, "in minimum time" or "while expending minimum energy."

The systematic approach to calculating an optimal control law begins with the choice of a *performance index* or *cost function*, a quantitative expression that can be used to determine how well a particular control scheme meets the design goals. Once a cost function is defined, the solution proceeds more or less automatically; therefore, it is wise to devote effort to the formulation of a cost function that accurately represents the goals and constraints of the problem at hand. It is usually simple to formulate a mathematical function that is positive definite and with a value related to a system's performance, for example, but the assignment of weighting coefficients and penalty terms requires greater insight into the problem. The definition of the performance index can also affect the ease with which a solution is obtained.

Let us for the moment consider a very general state-variable representation of an arbitrary system,

$$\dot{\mathbf{x}} = \mathbf{f}(\mathbf{x}, \mathbf{u}, t), \quad \mathbf{x}(t_0) = \mathbf{x}_0, \tag{B.40}$$

where, again, $\mathbf{x}$ is the system's state vector and $\mathbf{u}$ the control inputs. Our earlier matrix equation governing the state of an LTI system is a special case of eq. (B.40), and we shall return to it shortly. A correspondingly general performance index may be written

$$J = \int_{t_0}^{t_f} \ell(\mathbf{x}, \mathbf{u}, t)\, dt + m(\mathbf{x}_f), \tag{B.41}$$

where $\mathbf{x}_f = \mathbf{x}(t_f)$ and $\ell(\cdot)$ and $m(\cdot)$ are functions specified by the designer. It is often convenient to take the initial time t_0 to be zero; whether the final time t_f is finite or infinite has a more profound effect on the solution for the optimum $\mathbf{u}(t)$, and may sometimes be regarded as a design decision. Once these functions and parameters are specified, the problem becomes that of calculating the control $\mathbf{u}(t)$ that minimizes the cost function J over the interval $[t_0, t_f]$.

For a broad class of applications of practical interest, this optimization problem may be solved by using techniques based on the calculus of variations (Bryson and Ho, 1975; Stengel, 1994). The state equation in the form (B.40) is appended to the cost function as a constraint through the introduction of the vector $\mathbf{p} = (p_1 \ \cdots \ p_n)^T$ of Lagrangian multipliers, resulting in

$$\hat{J} = \int_{t_0}^{t_f} \left(\ell(\mathbf{x}, \mathbf{u}, t) + \mathbf{p}^T(t)[\mathbf{f}(\mathbf{x}, \mathbf{u}, t) - \dot{\mathbf{x}}(t)]\right) dt + m(\mathbf{x}_f). \tag{B.42}$$

The Lagrangian multipliers constitute the *costate*, and must in general be computed along with the state $\mathbf{x}$ as functions of time. The Hamiltonian, defined as

$$H(\mathbf{x}, \mathbf{u}, \mathbf{p}, t) = \ell(\mathbf{x}, \mathbf{u}, t) + \mathbf{p}^T(t)\mathbf{f}(\mathbf{x}, \mathbf{u}, t), \tag{B.43}$$

is now introduced and substituted into eq. (B.42). Following integration by parts,

$$\hat{J} = \int_{t_0}^{t_f} (H(\mathbf{x}, \mathbf{u}, \mathbf{p}, t) + \dot{\mathbf{p}}^T(t)\mathbf{x}(t))\, dt + m(\mathbf{x}_f) + \mathbf{p}_0^T \mathbf{x}_0 - \mathbf{p}_f^T \mathbf{x}_f. \tag{B.44}$$

The first variation of $\hat{J}$ is set equal to zero as a necessary condition for a minimum:

$$\delta\hat{J} = \int_{t_0}^{t_f} \left(\frac{\partial H}{\partial \mathbf{x}} \delta\mathbf{x} + \frac{\partial H}{\partial \mathbf{u}} \delta\mathbf{u} + \dot{\mathbf{p}}^T \delta\mathbf{x} \right) dt + \frac{\partial m}{\partial \mathbf{x}_f} \delta\mathbf{x}_f - \mathbf{p}_f^T \delta\mathbf{x}_f = 0. \tag{B.45}$$

For the equality to hold, the coefficients of $\delta\mathbf{x}$ and $\delta\mathbf{u}$ in the integrand must vanish, leading to a system of differential equations for the costate $\mathbf{p}$ and further conditions on the optimum $\mathbf{u} = \mathbf{u}(\mathbf{x}, \mathbf{p}, t)$. The terms outside the integral provide the boundary values $\mathbf{p}_f = \mathbf{p}(t_f)$.

To connect this to the state equation of the previous section,

$$\dot{\mathbf{x}} = \mathbf{A}\mathbf{x} + \mathbf{B}\mathbf{u}, \tag{B.46}$$

we consider a class of *linear quadratic* problems – so called because the plant equation is linear but the performance index is quadratic in the elements of $\mathbf{x}$ and $\mathbf{u}$. This relationship is conveniently expressed in terms of quadratic forms in the $\mathbf{x}$ and $\mathbf{u}$ vectors, as in

$$J = \frac{1}{2} \int_{t_0}^{t_f} (\mathbf{x}^T \mathbf{Q}(t)\mathbf{x} + \mathbf{u}^T \mathbf{R}(t)\mathbf{u})\, dt + \frac{1}{2} \mathbf{x}_f^T \mathbf{F} \mathbf{x}_f, \tag{B.47}$$

where the weighting matrices are all symmetric, and $\mathbf{Q}$ and $\mathbf{F}$ are positive semidefinite while $\mathbf{R}$ is positive definite. The Hamiltonian for this problem is

$$H = \frac{1}{2} \mathbf{x}^T \mathbf{Q} \mathbf{x} + \frac{1}{2} \mathbf{u}^T \mathbf{R} \mathbf{u} + \mathbf{p}^T (\mathbf{A}\mathbf{x} + \mathbf{B}\mathbf{u}) \tag{B.48}$$

and the optimum control turns out to be

$$\mathbf{u} = -\mathbf{R}^{-1} \mathbf{B}^T \mathbf{p}. \tag{B.49}$$

The costate can be computed from

$$\mathbf{p}(t) = \mathbf{K}(t)\mathbf{x}(t), \tag{B.50}$$

where $\mathbf{K}(t)$ is a symmetric matrix satisfying the differential Ricatti equation

$$\mathbf{K} = -\mathbf{K}\mathbf{A} - \mathbf{A}^T \mathbf{K} + \mathbf{K}\mathbf{B}\mathbf{R}^{-1}\mathbf{B}^T \mathbf{K} - \mathbf{Q} \tag{B.51}$$

subject to

$$\mathbf{K}(t_f) = \mathbf{F}. \tag{B.52}$$

This is not a simple equation, but its numerical solution is usually tractable.

An important version of this problem is that of the linear quadratic regulator (LQR), where the weighting matrices are constant and the final time $t_f = \infty$. Here

$$J = \int_0^\infty (\mathbf{x}^T \mathbf{Q} \mathbf{x} + \mathbf{u}^T \mathbf{R} \mathbf{u})\, dt \tag{B.53}$$

and the optimal control can be found to be

$$\mathbf{u} = -\mathbf{R}^{-1}\mathbf{B}^T\mathbf{K}\mathbf{x}, \tag{B.54}$$

where $\mathbf{K}$ is the unique positive definite solution of the algebraic Ricatti equation

$$\mathbf{0} = \mathbf{K}\mathbf{A} + \mathbf{A}^T\mathbf{K} - \mathbf{K}\mathbf{B}\mathbf{R}^{-1}\mathbf{B}^T\mathbf{K} + \mathbf{Q}. \tag{B.55}$$

The analysis, design, and simulation of LQR control systems is now commonly a feature of numerical software packages such as Matlab.

Although from the foregoing it may appear that optimal control is a forbiddingly complex subject, the basic goal of quantitatively comparing designs is fundamental to all areas of engineering. The cost functions and analytical approaches of optimal control are much broader than we've been able to sketch here, but much of the mathematics will likely be familiar from classical mechanics or from other optimization applications. If it is determined that active control is required for a structure to perform adequately, it is often worthwhile to undertake the design and implementation of an optimal control law rather than accepting the performance that can be achieved with a simpler, off-the-shelf controller.

B.6 Digital Control

So far, we have discussed plants and control systems whose inputs and outputs are smooth functions of time. Historically, much of control theory was developed in this setting, and many physical systems do, of course, function this way – structures being prime examples. On the other hand, there are numerous advantages to performing computations using digital rather than analog devices, and it is becoming ever more practical to use digital computers as components of active control systems. In this section, we introduce some of the terminology of discrete-time systems and indicate the relationships between analog and digital controls.

The most basic difference between continuous-time and discrete-time systems is that in the latter, signals are evaluated (sampled or computed) only at distinct instants of time. A variable $x(t)$ is represented by its values $x_k = x(t_k)$ at a sequence of times t_k, $k = 0, 1, 2, \ldots$. It may be regarded as constant (or otherwise simply interpolated) during the intervals between the times t_k, but for the purposes of analysis and design we are concerned only with the sequence x_k. Signals that are inherently continuous, such as the output of an analog sensor, can be discretized by an analog to digital converter (abbreviated ADC or A/D), while a continuous signal can be synthesized from a sequence of discrete values by a digital to analog converter (DAC or D/A). A common configuration in practice is the combination of a continuous-time plant, such as a structure, and a discrete-time controller, such as a digital computer. In this arrangement, the digital subsystem is preceded by an ADC and followed by a DAC; this is sometimes termed a sampled-data system. This allows the control design to be carried out largely in the discrete domain.

As the cost and complexity of digital hardware decrease, digital control systems are seeing wider applications. Two chief reasons for this popularity are stability (a numerical algorithm is immune to some of the phenomena, such as drift, that can plague analog circuits) and ease of modification. Once the cost of having a computer in the system is paid, additional complexity in the control algorithm itself is not prohibitive as it might be were the hardware not programmable.

The analysis and design techniques of digital control parallel both the classical and modern theories, in that one may work with either an input-output relationship or a state-space model. Because time and all the variables that depend on it are treated as sequences of discrete values, what were previously differential equations are now difference equations, and solutions, rather than being continuous functions of time t, may be described by recurrence relations in terms of the index k. The Laplace transform is not applicable in this context, but can be replaced for most purposes by the z-transform (in which integration is replaced by summation). Block diagrams can be manipulated much as for continuous-time systems, characteristic equations can be used to test stability, and so forth.

Because the mathematics of sequences is likely to be unfamiliar to many readers, we shall not go into detail on these topics here. However, some points should be emphasized as they pertain especially to structural control. Structures tend to have "fast" dynamics compared to many other engineering systems (e.g., the vibration of a spring-mass system typically occurs at much higher frequency than the variation of concentrations within a chemical reactor). The frequency at which a control system must respond places a lower bound on the frequency at which continuous signals must be digitized (the sample rate). If a signal $x(t)$ is not sampled at a rate adequate to capture its high-frequency content, the resulting series x_k is not a faithful representation of the original function. Even when the sample rate is adequate, one must be aware of the time delays in the ADC, computation, and DAC steps. These may be brief, but they are often constant and therefore lead to increasing phase delays as the frequency of operation increases. Such considerations do not necessarily argue against the use of digital controllers in smart structures, but they do demonstrate issues that might not arise in other applications but can be critically important in the present context.

BIBLIOGRAPHY

Bathe, K.-J. 1982. *Finite Element Procedures in Engineering Analysis*. Englewood Cliffs, NJ: Prentice-Hall, Inc.

Bryson, A. E., Jr. and Yu-Chi Ho. 1975. *Applied Optimal Control*. New York: Hemisphere Publishing Corporation.

Chen, C.-T. 1984. *Linear System Theory and Design*. New York: Holt, Rhinehart and Winston.

Franklin, G. F., and J. D. Powell. 1980. *Digital Control of Dynamic Systems*. Reading, MA: Addison-Wesley.

Friedland, B. 1986. *Advanced Control System Design*. Englewood Cliffs, NJ: Prentice Hall.

Gabel, R. A., and R. A. Roberts. 1980. *Signals and Linear Systems*. 2d ed. New York: John Wiley & Sons, Inc.

Inman, D. J. 1989. *Vibration: With Control, Measurement and Stability*. Englewood Cliffs, NJ: Prentice Hall.

Kuo, B. C. 1980. *Digital Control Systems*. New York: Holt, Rhinehart and Winston.

Kuo, B. C. 1995. *Automatic Control Systems*. 7th ed. Englewood Cliffs, NJ: Prentice Hall.

Skelton, R. E. 1988. *Dynamic Systems Control: Linear Systems Analysis and Synthesis*. New York: John Wiley & Sons.

Stengel, R. F. 1994. *Optimal Control and Estimation*. New York: Dover Publications, Inc.

Weinstock, R. 1974. *Calculus of Variations: With Applications to Physics and Engineering*. New York: Dover Publications, Inc.

Index